The Pueblo Revolt and the Mythology of Conquest

The Pueblo Revolt and the Mythology of Conquest

An Indigenous Archaeology of Contact

Michael V. Wilcox

UNIVERSITY OF CALIFORNIA PRESS
Berkeley / Los Angeles / London

University of California Press, one of the most distinguished university presses in the United States, enriches lives around the world by advancing scholarship in the humanities, social sciences, and natural sciences. Its activities are supported by the UC Press Foundation and by philanthropic contributions from individuals and institutions. For more information, visit www.ucpress.edu.

University of California Press
Berkeley and Los Angeles, California

University of California Press, Ltd.
London, England

Library of Congress Cataloging-in-Publication Data

Wilcox, Michael V. (Michael Vincent), 1967–
The Pueblo Revolt and the mythology of conquest : an indigenous archaeology of contact / Michael V. Wilcox.
p. cm.
Includes bibliographical references and index.
ISBN 978-0-520-25205-9 (alk. paper)
1. Pueblo Revolt, 1680. 2. Pueblo Indians—Colonization. 3. Pueblo Indians—Government relations. 4. Violence—Political aspects—Southwest, New—History. 5. Spaniards—Southwest, New—History. 6. Southwest, New—Race relations. 7. Southwest, New—History—To 1848. 8. Pueblo Indians—Antiquities. 9. Ethnoarchaeology—Southwest, New. 10. Pueblo Indians—Historiography. I. Title.

E99.P9W55 2010
978.9'02—dc22 2009023615

Manufactured in the United States of America

16 15 14 13 12 11 10 09
10 9 8 7 6 5 4 3 2 1

The paper used in this publication meets the minimum requirements of ANSI/NISO Z39.48–1992 (R 1997) (*Permanence of Paper*).

Cover illustration: The Jemez Mission of Guisewa. Photograph by the author.

For Julie, Hannah, Joslin, Jake, and Kian

Contents

communities, is very real. And the kinds of questions archaeologists ask (or do not ask) have very real political and material consequences for contemporary Native Americans. The power to create narratives of the past, to tell some stories, to articulate some histories and not others, is and has always been political.

In working with those remains, I realized that in siding and sexing the bones, looking for paleopathologies, determining age, evaluating peri- and postmortem fractures, and so forth, I was inscribing these people with a history of my own making. Wrapped in the language of medicine, I was writing a postmortem or autopsy a thousand years after death. But to what ends? Where would this information go? Who would read it? Was this new narrative, these new inscriptions of meaning, the truth? Or did this type of recognition of the body obscure other worthwhile, more prescient, and valuable historical narratives? Why was archaeology so comfortable in privileging the former and so disinterested of the historical?

These questions have served as the basis for both this book and the development of what I have termed "Indigenous archaeology." This book is my first attempt at this unique approach to the past. Despite the fears of some archaeologists who may wish to impose a kind of intellectual apartheid, a "separate yet equal" partition of archaeological research (McGhee 2009), an Indigenous approach to archaeology does not limit or displace scientific research. It merely poses as its fundamental project the reintegration of Indigenous materials, remains, history, and research with contemporary Indigenous peoples. This work, currently practiced on a large scale throughout the United States under the auspices of the Native American Graves Protection and Repatriation Act (NAGPRA), has remained terribly underrepresented in North American archaeological journals, is poorly understood by many people who do archaeology in the United States, and has been criticized by a vocal, conservative group of scholars who refer to themselves as "scientific archaeologists." These individuals have not welcomed the changes ushered in by NAGPRA and see the entry of Native Americans (myself and others) into the discipline as a threat to the future of the field.

No such threat exists. It is my belief that the future intellectual and ethical vitality of the discipline depends on collaborative methods (Colwell-Chanthaphonh and Ferguson 2008; Silliman 2008), the reversal of terminal narratives that explain the disappearance and alienation of the past from Native peoples, and the rejection of the idea that Indian scholars have no place within the discipline of archaeology. We do have a place, but one should not expect us to create the same kinds of archaeologies that existed before our entry into the field. To expect so is to embrace the worst of all kind of tokenism. No field that calls itself a science should be afraid of new information, new perspectives, or new interpretations.

Acknowledgments

I wish to thank the small but growing group of Indigenous archaeologists—Sonya Atalay, Dorothy Lippert, Joe Watkins, Sven Haakenson, and Desiree Martinez—who have supported this work and are at the forefront of the Indigenous archaeology movement. People such as Larry Zimmerman, Randy McGuire, Bob Preucel, T. J. Ferguson, Roger Anyon, Kent Lightfoot, Steve Silliman, Chip Colwell-Chanthaphonh, and Barbara Mills have been advocating for the inclusion of Indigenous perspectives in archaeology for years. They might not consider themselves as such, but I believe that each of these scholars practices a kind of Indigenous archaeology and, in some cases, has been doing so for quite some time. I have drawn considerable inspiration from their work. Similarly, David Hurst-Thomas has been a very influential scholar in his approach to Indigenous peoples and the past. I have always been amazed by the range and depth of his scholarship and interests. His *Columbian Consequences* set is among the most widely cited works in the history of contact, and I have appreciated his feedback whenever he has offered it. Judith Bense, Kathleen Deagan, and Anne Ramenofsky have been similarly influential albeit as authors of works I respect but sometimes have disagreed with.

Closer to home, I have benefited greatly from the community of archaeologists at Stanford University as well as at the University of California, Berkeley. My colleagues at Stanford—Ian Hodder, Lynn Meskell, and Barbara Voss—have done far more to help me than they are aware of. I would not have been able to write this book without their intellectual and scholarly input and support. The Department of Cultural and Social Anthropology at Stanford has been my home for the past seven years. Jim Ferguson, Paulla Ebron, Liisa Malkki, Sylvia Yanagisako, and Miyako Inoue have been the best of colleagues for the past few years. At Berkeley, Meg Conkey and Kent Lightfoot, along with their many talented students, have provided a supportive community throughout the writing of this book. Kent in particular has been a huge help and influence in my own work. He was my faculty mentor during the 2004–2005 school year while I was a National Academies of Sciences Ford Fellow at Berkeley. Carole Mandryk, Bill Fash, and Bob Preucel all served as faculty advisers while I was a graduate student at Harvard. I could not have begun this research without their help. The entire project would be impossible without the help and guidance of Bob Preucel at The University of Pennsylvania. The Kotyiti Research Project was his idea. He showed me how to do collaborative archaeology by his actions—this at a time when few archaeologists had recognized that the field had changed with NAGPRA. Bob is and has always been an inspiration to me.

John Ware at the Amerind Foundation has facilitated some of the most exciting research in archaeology through his seminar series. Matt Veirling of Mattsmark.com has been a fantastic cartographer and graphic artist. Richard Flint and Matt Smaeder have been helpful in clarifying some of the sites related to the Coronado entrada. I appreciate their scholarship and helpfulness. Most important, I wish to thank the people of Cochiti Pueblo: Governors Henry and Joseph Suina, and Governor Trujillo each played an important part in telling this story. April Trujillo, Jeff Suina, and Christine Suina deserve special thanks, as do the Montoya family. Their hospitality and humor during the long days of the summer field seasons are a special memory of mine. Finally, I wish to thank my wife Julie for all of her help, and likewise to Uncle Greg Jennings.

CHAPTER I

Repatriating History

Indigenous Archaeology and the Pueblo Revolt of 1680

Cochiti, New Mexico

Every July, on the Feast of San Buenaventura, an eclectic mix of tourists, spectators, and neighbors filter through the dusty tangle of lanes in the old village and gather on the periphery of the main plaza at Cochiti. To the east stands the Stalinesque monstrosity known as Cochiti Dam. Designed in the 1960s in an effort to protect the people of Cochiti (and Albuquerque) from a five-hundred-year flood, the Army Corps of Engineers shoved the sixty-five million cubic yards of earth and rock into a looming, black pile of vesicular basalt (figures 1 and 2). The people of the Pueblo fought the project from beginning to end, watching as the water table in the village rose each year. The leaky, black mountain lets the water through, but it keeps the rising sun from shining on the front doors of the mission until the morning services are completed. Following the morning mass, a small group of men carry the wood and plaster image of the patron of the Pueblo from the altar of the church and into the consecrated ground that lies between the mission and the Rio Grande.

Only members of the pueblo are invited into the mission. Spectators stand clear of the low-walled *campo santo*, or graveyard, and wait while the wooden likeness of San Bonaventura is conveyed in silence to the main plaza. There, a *ramada* of evergreen boughs shades a collection of trophy elk heads taken in the fall, woven textiles, candles, and other offerings. Dressed in the heavy brown robes of the Franciscan Order is a white-haired priest seated alongside an assortment of elders from the Pueblo.

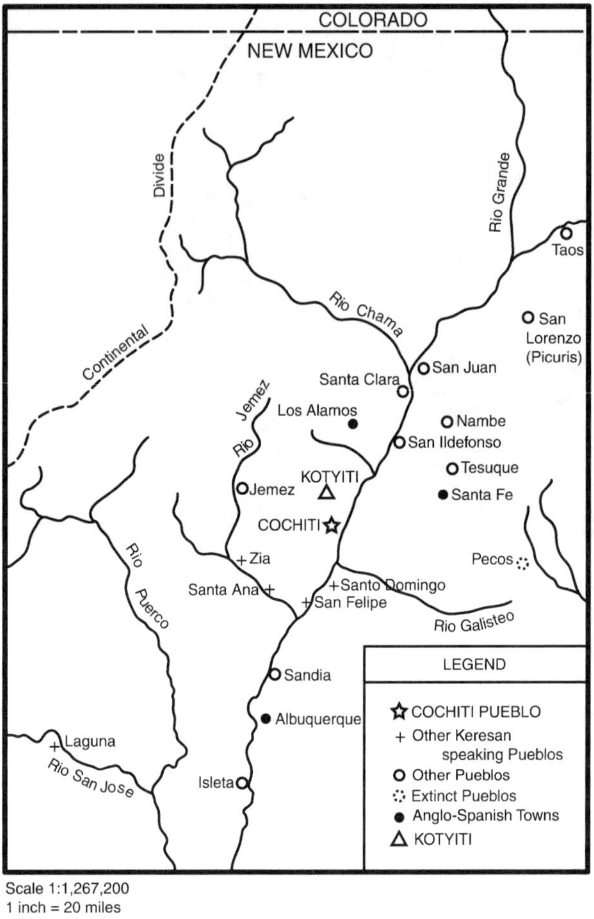

FIGURE 3. Location of Old Cochiti.

hundreds of miles, even beyond the modern sprawl of Albuquerque (figures 3, 4, and 5). Further up the mesa, surrounded on three sides by the sheer cliffs two hundred meters off the valley floor are the remains of a larger, rectangular village known as LA 295. Divided in two by a small central line of rooms, the village has two kivas, one in each plaza. One has a tall spruce growing from the center. Some of the outer walls are still standing. The reddened plaster is burned by a fire and from the size and arrangement of stones, I can see that these must have formed a defensive outer wall separating the pueblos from the violent assaults of the Juan Diego de Vargas during the "bloodless reconquest" of April 1694.

FIGURE 4. Colonial New Mexico and the Borderlands of New Spain.

Selected by the Pueblo for our project by the governor, the high school–aged interns are reticent and guarded—typical of most kids that age—but seem to be struggling to figure out whether their selection for the project is privilege or punishment. They smile when we refer to the village as "Hanut Koyiti," the orthographic rendering applied to the site by an earlier generation of anthropologists trained to read and write in our cryptic ancestral alphabet. A rough translation would be "Moon House," but the interns are reluctant to tell us the correct pronunciation for the village or any of the other landmarks visible from the mesa. We fumble with the outdated site maps copied from the century-old notes of Adolph Bandelier and point out pottery types and styles to the

FIGURE 5. The View from Cochiti Mesa.

politely bored interns. Before visiting Cochiti and then the site in the 1880s, Bandelier had been removed from the neighboring Pueblo of Santo Domingo after ignoring his guests' admonitions against viewing religious ceremonies and taking photographs. He later used his experiences as the inspiration for his classic study of Pueblo clowns called "the Delight Makers."

But the village holds a much larger meaning in the history of the Pueblo people, and we are aware of this significance as we tread carefully around the perimeter of the melting adobe and stone walls. It was here in the 1680 that the leaders of the Pueblo Revolt gathered, planned, and executed the most successful Indigenous rebellion in North America. For just over a decade, the Pueblo world was free from crippling tribute demands, religious oppression, and enslavement, all enforced with the peculiar "legal and rational" forms of social violence negotiated between the Spanish Crown and the popes in the administration of Indians. The brutality of these encounters, here at the extreme edge of empire, was no fiction. The hundreds of letters, internal documents, reports, and commentaries of Spanish officials provide a damning testimony of Spanish brutality between 1540 and 1680 on New Spain's northern frontier. They were not written as plaintive protest essays (Bartolome de Las

Casas's *Brevísima relación de la destrucción de las Indias* [1552] was published thirty-three years after Cortés and twelve years after Coronado's entry into New Mexico) but describe in detail the actions, justifications, and personal reflections of the Spaniards in their own words. The colonial ventures of English settlement in the seventeenth century were equally brutal, but locked in the internal communiqués of some of the world's first transnational corporations, the moral justifications for both slavery and the forcible removal of Indians were never seriously debated. Spain, as the military branch of the Catholic Church, held its colonists to much higher and more rigorous standards of behavior, and the documents reveal the tensions between the parallel and contradictory motives of salvation and exploitation. None of the interns has read or even heard that these documents exist.

In an ironic (yet predictable) feat of colonial revisionism, the pueblos have been largely characterized as aggressors in both the history texts and within the monuments in the plaza at Santa Fe. Manufactured by Euro-American historians and upheld by New Mexico's founding fathers, these colonial narratives were (and to some extent still are) enforced by the power of the state. In classrooms and public spaces, the ability to imagine and defend "the Narrative" has both material dimensions and political consequences. In the 1970s, an unknown vandal exacted his own revisionism by jack-hammering the words "Savage Indians" off a granite obelisk commemorating New Mexico's Indian Wars (figure 6a and 6b). Parallel narratives, dissonant in meaning, each reflecting and enforcing a segregated system of knowledge and power, meaning and interpretation, identity and ethnicity.

For the Pueblos, the message of the dance and the struggle over narrative is encoded in the small, nondescript arrangement of rooms, plazas, and kivas on the prow of this mesa. This material world, represented (or not represented) through the disciplinary filters peculiar to American archaeology, has a powerful symbolic and political significance to the Pueblos. The surprised looks of the interns when we refer to villages and landmarks with the Laboratory of Anthropology (LA 123) system of classification or give names to common objects they have seen and lived among their whole lives (biscuit wares, Jemez black-on-white lithic assemblages) reveal a sense of unease and a recognition that their material world—indeed, their own histories—have been appropriated, bounded by a system of knowledge in which the Pueblos themselves have become outsiders and aliens. The anonymity of the village reflects a troubling disassociation between contemporary Native peoples and

FIGURE 6A. Monument to Indian Wars, Santa Fe Plaza.

the scholarly disciplines that interpret and represent the materials, cultures, and histories of Indigenous peoples to the general public. For most of the past century, archaeologists, historians, and anthropologists have had free reign in the interpretation of Indian histories; they have been able to impose a variety of scientific, legal, academic and professional barriers between themselves and descendant communities of Native Americans. But that relationship is changing.

In September 2005 at the Capitol Building in Washington, DC, the hundredth and final entry to National Statutory Hall was made by the State of New Mexico. The marble likeness, carved by Jemez Pueblo sculptor Cliff Fragua, depicts Po'pay, a Tewa religious leader from the Pueblo of San Juan. Po'pay joined Cherokee leader Sequoyah as one of two Native Americans in the collection. His presence mirrors a significant realignment in the historical narrative of colonization and the location of Native Americans within that narrative. It also reflects a resurgence in sovereignty and political visibility for Indigenous peoples. With the material assertion of the statue, the pueblos have enacted a social and symbolic

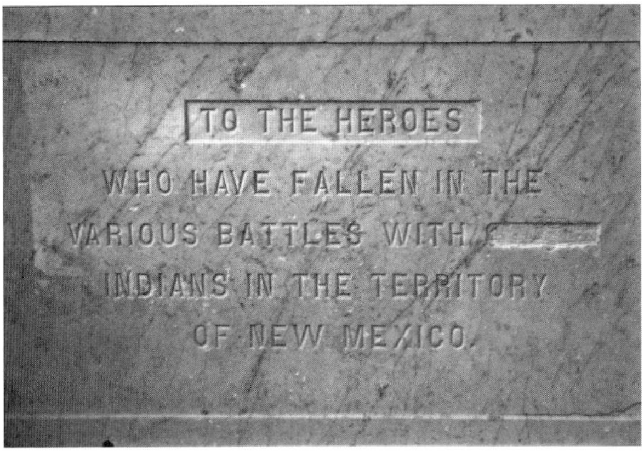

FIGURE 6B. Monument to Indian Wars, Santa Fe, Inscription.

presence within contemporary American society and demanded a greater voice in the representation of the material remains of their ancestors.

Unfortunately, the discipline of archaeology was unable to anticipate this tectonic shift and reluctant to acknowledge the ways in which *contemporary* archaeology had negatively impacted Native Americans. As a consequence, by an act of Congress in the early 1990s, the leaders of the Society for American Archaeology (SAA) were forced into negotiating for access to what they had formerly considered to be their own exclusive *data set.* Until then, access to artifacts and human remains, excavation and site reports, and plans for research were restricted to a small group of museum workers and archaeologists.

Passed in 1990, the Native American Graves Protection and Repatriation Act (NAGPRA) was a watershed moment for both archaeologists and Indigenous peoples. For Native Americans, NAGPRA was a political victory with roots in the civil rights movements of the 1960s. Following an organizational script borrowed from other ethnic coalitions, a small group of politically savvy, highly motivated, and well-networked activists were able to convince the U.S. Congress and a conservative president (George H. W. Bush) to pass the most extensive and transformative legislation affecting Native Americans since Franklin Roosevelt's Indian New Deal in the 1930s.

Framed as a piece of human rights legislation and written in anticipation of the five-hundred-year anniversary of European colonization of the Americas, NAGPRA marked a period of critical reflection and reconciliation among colonial governments and religious officials about the historical legacy and contemporary meaning of colonization. With the stroke of a pen, controls over human remains, grave goods, and objects of cultural patrimony were transferred from museums, universities, and archaeological facilities to local Indian communities.

But the transfer of power was more than material. In articulating a common historical experience as the subjects of European colonization, NAGPRA drew power from a global Indigenous narrative and transformed that common experience into political action at the local level. In requiring museums and archaeologists to seek connections between the artifacts, sites, and materials in their collections and *living* Indigenous groups, Native Americans were both asserting their presence within a contemporary political and cultural landscape and *demanding a new narrative that explained that presence.*

The resulting dialogue between contemporary Native Peoples and archaeologists has dramatically transformed the discipline. While cultural and social anthropology faced representational critiques levied by Indigenous peoples in the years following decolonization (ca. 1945–1960), descendant communities of Native Americans were effectively walled off from archaeological scholarship by the positivist stance of processual archaeology. The material substances of archaeology—sites, features, artifacts, and human remains—didn't "talk back"; the interpretations of these materials had never required that archaeologists speak to anyone, except each other. Nor were archaeologists required to submit the human subject protocols that strictly governed the activities of psychologists, sociologists, and anthropologists. Because the dialogue with Native Americans was not motivated by any substantial internal theoretical critique, NAGPRA forced archaeologists to confront some of the most challenging issues in the history of the discipline:

- How did the interests, objectives, and meanings of archaeological practice come to diverge so sharply with those of Native Americans?
- What responsibilities did archaeologists have in relation contemporary Indian peoples?
- How had the narratives of contact, conquest, and colonization developed by anthropologists, historians, and archaeologists so alienated living Native Americans? Did any of these narratives explain the persistence of Indigenous traditions, religions and languages embodied in the Feast Day Dance at Cochiti or the statue of Po'pay in 2005?

Given the lack of contact between contemporary Native communities and archaeologists, the persistence of the perpetually vanishing Indian, most recently incarnated in Jared Diamond's narrative of accidental conquest *(Guns, Germs, and Steel)* is not surprising. But what if the positions and assumptions of the scholars were reversed? What if archaeologists were asked to explain the continued presence of descendant communities five hundred years after Columbus instead of their disappearance or marginality? Moving beyond the largely empty and gestural process of compliance negotiations, this work brings traditional studies of contact and colonization to a new level of engagement with contemporary Native Americans and their histories. Using the Pueblo Revolt of 1680 as a case study, this work provides an explanation for the persistence of contemporary Indian peoples as well as a model for the application of Indigenous archaeology in North America.

Combining the interdisciplinary nature of early colonial or contact period archaeology with the critical evaluations of postcolonial studies, this book demonstrates the ways in which an interdisciplinary scholarship, engaged in critical readings of primary historical documents and contemporary narratives of European colonization, can enable a transformative theoretical movement within the fields of archaeology, anthropology, and colonial history. This involves the active dismantling of *terminal narratives*—accounts of Indian histories which explain the absence, cultural death, or disappearance of Indigenous peoples. Terminal narratives have had a profoundly damaging effect on popular conceptions of contemporary Indigenous peoples. Most of the information communicated to the general public about Indigenous peoples is associated with destruction and disappearance. Conquest narratives, disease, cultural and demographic collapse, acculturation, and assimilation are all closely associated with Indigenous peoples. Writing against the essentialist theories of acculturation generated by anthropologists as well as the narratives of disease and abandonment deployed by archaeologists, I argue for a reformulation of traditional approaches to contact period or borderlands studies and colonial history.

Modeled in such a way, Indigenous archaeology represents a methodological and theoretical commitment to the reintegration of archaeological materials and contemporary Indigenous peoples. In reengaging the fields of archaeology, anthropology, and history with the histories and perspectives of contemporary Native Americans, Indigenous archaeology moves these disciplines beyond the often intransigent and static formulations imposed by the segregation of Indian and European historical narratives and initiates a new series of

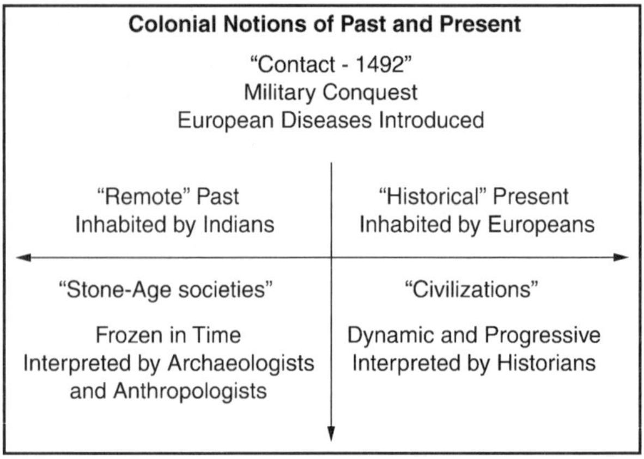

FIGURE 7. Professional Domains of Contact Period Scholarship.

The statements of Jared Diamond and the late anthropologist Eric Wolf reflect a persistent theme in the archaeology and history of the contact period. For many archaeologists and colonial historians, the line between history and prehistory divides more than the professional domains between these fields. First contacts signal the demise of Native cultures, traditions, and communities through disease, conquest, and acculturation (figure 7). These divisions reflect and reproduce more than the separate professional, methodological, and theoretical domains of archaeologists and historians. For different reasons, neither archaeologists nor historians have adequately examined or been particularly interested in explaining the *persistence* of Indigenous cultures following contact. Anthropological and archaeological studies of "pristine" Indigenous cultures seem to end, along with their subject matter, where colonial history begins. As a result, archaeological studies of the contact period tend to explain the disappearance of Native communities through either the spread of European pathogens or, for historical archaeologists, the spread and development of European settlements.

Diamond's quote reflects a prominent theme among these scholars. Generally, historians have emphasized the period of contact as a historical moment in which the pre-Columbian or Indigenous *past* is segregated professionally and theoretically from the advent of Western history. The practical result of these professional divisions is that Indians effectively disappear when archaeological investigations end and historical studies begin. Archaeological studies of the contact period in the

Southwest have only reinforced this perception. In the search for nomothetic models of human behavior, early practitioners of the New Archaeology used the ethnological studies of their intellectual ancestors as a foil. The scientific study of the past conceived of what had been Indian prehistory as the data set of a more remote past. From this perspective, one could argue that the advent of the New Archaeology marked the end of the study of Native histories.

This raises some important questions. First, how real is this dichotomy? To what degree has the scholarship of archaeologists, anthropologists, and historians in North America unwittingly submerged the history of colonized peoples? If we follow the interpretations of Diamond, how do we explain the persistence of indigenous cultures and traditions *following* contact? Most important, if Native populations were decimated by diseases supposedly introduced 150 years earlier, how do we interpret the persistent efforts of the Pueblos to resist Spanish domination or the successful expulsion of colonists following the Revolt of 1680?

FROM COLONIAL TO POSTCOLONIAL TO INDIGENOUS ARCHAEOLOGY

As Eric Wolf demonstrated in *Europe and the People without History* (1982), the study of history has traditionally involved a genealogical analysis of European nation-states as both discrete and dynamic units of analysis.[1] Europe is conceived of as the center from which global changes take place, transforming the *traditional* or *static* societies of Asia, Africa, and the Americas through contact, conquest, and assimilation. In *People without History*, Wolf went so far as to question the label *ethnohistory*, suggesting that the separation of *primitive* and *civilized* histories marginalized the perspectives of colonized peoples for economic, political or ideological reasons (Wolf 1982: 18–19). Responding to criticisms that the "objective" work of anthropologists had at times been deployed in support of colonial regimes and was compromised by the power relations between colonial researchers and colonized peoples, Wolf suggested that the task of postcolonial historians and archaeologists should be the *recovery* of the largely unwritten, unarticulated history of colonized peoples using an interdisciplinary methodology:

> It is only when we integrate our different kinds of knowledge that the people without history emerge as actors in their own right. When we parcel them out among the several disciplines, we render them invisible—their story which is our story, vanishes from sight.[2]

In North America, Wolf's comments have a particular resonance, especially considering the lack of communication between archaeologists and Native peoples prior to the passage of NAGPRA. While we have witnessed a great expansion in the popularity and contributions of processual archaeology in the Southwest over the past forty years, the study of postcontact, historical interaction between Spanish colonists and the Native peoples and the use of ethnographic and historical texts have been largely abandoned as a viable and potentially informative approach to understanding social changes on a grand scale. Fortunately, with the passage of NAGPRA and the integration of Native American perspectives, this trend is changing.[3] The emphasis on the scientific analysis of archaeological sites has generally stimulated an aversion to the use of history in the study of Native American material culture; the assumption seems to be that the investigation of *process* and *history* are separate areas of study. While processual archaeology shied away from historical studies of Native peoples, social and cultural anthropologists began to see the colonial expansions of early modernity as a legitimate and prolific field of inquiry. Among Wolf's many contributions was his application of conflict oriented, Marxist social theories to anthropological studies. In an increasingly global, postcolonial world, the analysis of *conflict* and *resistance* as social processes has proven to be of great value to those interested in the comparative study of human societies.

Wolf's insistence that Western scholarship had helped render Native Americans and Africans *"a people without history"* foreshadowed the development of what has come to be known as *postcolonial studies.* Defined as a "discourse of opposition which colonization brings into being" (Ashcroft et al. 1995: 117), postcolonialism encompasses a multitude of critical approaches to both the subject and practice of colonization. In delving into the processes of colonization, nationalism, and ethnogenesis, as well as the role of Western scholarship in the representation of colonized peoples, postcolonial approaches attempt to articulate the perspectives and resistance strategies of colonial subjects. In deconstructing the monolithic, Eurocentric master narratives of colonial histories, postcolonial scholars such as Edward Said (1978), Homi Bhabha (1990), and Gayatri Spivak (1996) shifted the focus of debate and discussion to the complex interplay of myth, symbol, and memory in the construction of indigenous and national ethnic identities. Located within this context, events such as the Revolt of 1680 offer a unique opportunity to reevaluate traditional narratives of conquest, subordination, and Spanish colonialism.

Through an unintentional theoretical bias against postcontact studies of Puebloan peoples, the unique perspectives of Native peoples have in many cases remained, in Wolf's words, "unarticulated." However, if we shift our perspectives, as Wolf suggests, and attempt to integrate the histories of Native peoples and Europeans, if we attempt to explain the continued presence of Puebloan communities, cultures, and traditions as equal partners in the present, then we are led to examine a separate set of historical processes, events, and patterns of accommodation and resistance.

Rethinking Acculturation and Demographic Collapse: Puebloan Resistance and Rebellion in Context

One consequence of the dominance of the dominance of "disease and acculturation models" of the postcontact period has been a lack of scholarly attention paid to the subjects of conflict, violence, and resistance between colonists and Native peoples through extended periods of time. Spain, like most colonial regimes, routinely deployed violence as a means of establishing and maintaining dominant social, economic, and political relationships among colonized populations. Unfortunately, most Borderlands historians, many of whom were trained in the bardic tradition of George Bancroft and Herbert Eugene Bolton, have been reluctant to address the tactical significance or moral, legal, and religious justifications for the use of violence by Spanish adventurers and religious and secular officials. Wary of the legacy of Black Legend depictions of Spanish barbarism and eager to elevate the colonial history of Hispanic Americans in relation to Anglophile studies of the eastern colonies, Borderlands historians rendered a largely celebratory, "bloodless" version of Spanish colonization in the Southwest.[4]

Among historians of New Mexico, from Charles Lummis to David Weber, Coronado is memorialized as a heroic and adventuresome pioneer of Western civilization in the tradition of Francisco Pizarro and Hernán Cortés. In many of these studies, the establishment of Spanish culture and religious and political institutions and the dissolution of Native cultures and history seems to hinge on a series of unique and momentous historical moments. But in emphasizing the triumphant moments of conquest of Coronado, Pizarro, and Cortés popularized by historians, we miss the diversity of responses to among different Native communities. For many Native peoples, the processes of contact and subordination and the long-term animosities generated by colonial interaction generated creative responses of resistance and accommodation.

An examination of the historical records of colonial administrators in New Spain reveals that indigenous movements and rebellions continually plagued colonial officials well into the twentieth century. In nearly every area of Spain's colonial empire, the conquest and subjugation of Native peoples was neither instantaneous nor bloodless. The Inca Rebellion in 1534 and the Tupak Amaru movements of 1572 and 1779 helped shape the relations between Indian peoples and Spanish colonists in Peru. In Mesoamerica, the defiance of the Yucatán Maya between 1517 and 1534 led to the Great Revolts of the Maya in 1546 and 1761. These movements of resistance were followed by the Yucatán Caste War and Cults of the Talking Cross of the nineteenth century. Stalling the consolidation of the northern territories of New Spain, colonial and religious authorities were challenged during the Mixtón and Tarahumara Rebellions of the sixteenth to nineteenth centuries. Through these and other movements, Native peoples actively confronted the efforts of colonists to extract tribute and extinguish Native traditions and religious practices. While the initial conquests of large pre-Columbian empires allowed colonists to control large Indigenous populations, extract tribute, and more closely monitor and transform Indigenous religions and cultural practices, the colonization of New Spain's northern and southern frontiers was a much more costly and protracted process. For scholars of the contact period, this point raises several important questions:

- How do we explain the phenomena of Indigenous rebellions throughout the course of Spanish colonial history? What exactly *is* a conquest?
- Does military action and social violence actually create and activate fundamentalist movements that are inherently insular? Does acculturation ever really happen in these contexts?
- How do we account for the persistence of Indigenous traditions today? Are there any narratives that explain the presence of Indigenous peoples?

There is no doubt that contact between the Old World and the New World transformed the cultural, religious, economic, and social histories of both European and Indigenous peoples. But Native peoples had developed strategies to accommodate and coexist with the cultural practices of alien groups long before European contact. The idea of a monolithic pre-Columbian social universe is an important element in of the mythology of Western consciousness. It is also utterly fictional. In fact, "contacts" and conflicts are processes that all societies experience. All societies have some means of regulating interaction between groups, segregating certain behaviors and resisting the influences of other more complex and powerful neighbors. The study of the processes of contact and interaction between cultures, and the maintenance of cultural differences, is the subject of *ethnicity*—a theoretical domain and analytical tool that is absolutely essential in deconstructing Western mythologies of primitivity and the development of an Indigenous archaeology.

During the Revolt of 1680, several "refugee sites" were created unifying distinct and disparate linguistic and religious communities in an effort to expel the Spanish and reassert traditional religious and social practices (figure 8). Historic period regional demographic shifts have traditionally been interpreted using a combination of acculturation based on catastrophic, disease-population crash models. For the most part, the processes that lead to ethnic conflicts (including acts of violence, social and economic subordination, and religious suppression) have not been accounted for in many historical or archaeological contact period studies. While regional abandonments can be interpreted as resulting from catastrophic epidemics that devastate entire communities, population shifts during the Revolt of 1680 were motivated by a desire to maintain social distances and restore a traditional social and religious order. As such, these abandonments imply a very different set of motives and social consequences for Puebloan people.

In examining the responses of Puebloan people to the violent encounters of the sixteenth-century *entradas* (literally, "entrances," or expeditions), we find that abandonment and mobility were an important means of preserving and protecting Pueblo traditions and communities. In many cases, abandonment signaled not the end of communal life but an active and conscious effort to preserve local authority, communally based status systems, and familial, clan, and religious identities.

For archaeologists, the study and documentation of Revolt period communities offers a unique opportunity to examine the relationship between social conflicts, regional abandonments, and the settlement of new communities. While disease and acculturation were obvious factors in the interaction of Pueblos and Spanish colonists, I suggest that a more profitable analysis of contact period social interaction should involve an analysis of conflict and resistance. In applying the models of ethnicity and ethnic conflict to the colonial period, I hope to refine our

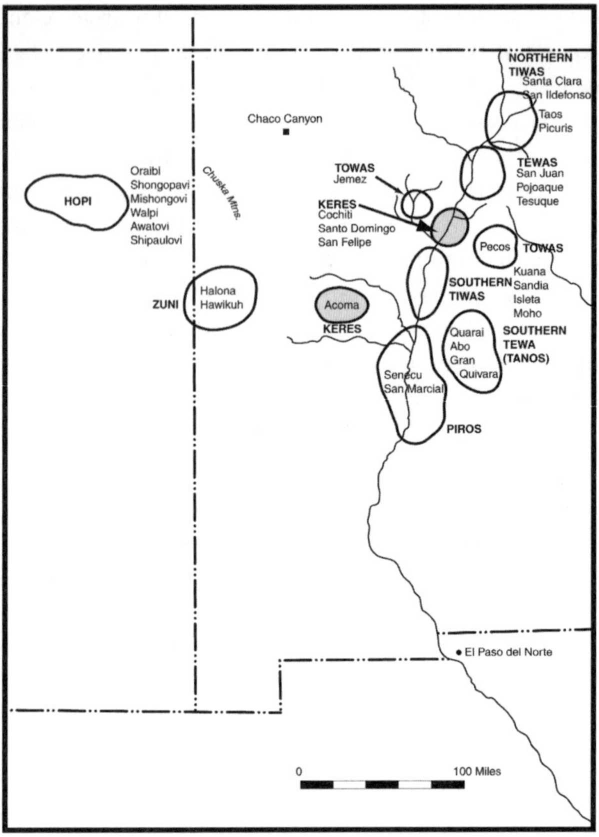

FIGURE 8. Pueblo Communities of the Upper Rio Grande.

understanding of early colonial interaction and, following the interdisciplinary methods of Eric Wolf, help integrate Pueblo and colonial histories and explain the persistence of Puebloan cultures and traditions.

Toward the Development of an Indigenous Archaeology

Published in 2000, Native American archaeologist Joe Watkins's *Indigenous Archaeology* identified some of the central tensions and theoretical questions associated with the development of a post-NAGPRA

The ideas that pre-Columbian, Indigenous communities lacked an awareness of social differentiation from other groups and that prior to the arrival of Europeans contacts with other alien communities did not occur are both implicit assumptions in "contact period" studies. As Wolf suggests, "the idea of the "pristine primitive" implies a static, essentialized view of Indian communities. By convention, the *contact period* commonly refers to the temporal range in which European colonists and Native peoples established social relations.

Area of Study

While this study attempts to situate conflicts and the Revolt of 1680 with reference to regional and global historical processes of the sixteenth and seventeenth centuries, the primary geographic focus of archaeological research is limited to the Upper Rio Grande (see figure 8). The general region of the Pueblos is defined by Cordell (1984). The regional area known as the Pajarito Plateau, defined first by Hewett (1938), serves as the setting for the settlement pattern study in the second half of the text.

Organization of the Book

This book is organized in three general sections. The first three deal with Indigenous archaeology, the creation of the invisible Indian, and a discussion of ethnicity theory. Each provides a model and methodology for reconnecting Indians with the past and dismantling the mythology of the perpetually vanishing Indian. This section forms the basic theoretical framework of (my version of) an Indigenous archaeology. The next section is more historically oriented and uses the methods developed in the first section in a reexamination of primary historical materials related to Spanish politics, militant Christianity, and the motives behind the slave trade and entradas of the sixteenth century. The final two chapters integrate historical documents and archaeological data in a more focused study of the archaeology of the Pueblo Revolt.

CHAPTER SUMMARIES

Chapter 2, "Creating the Invisible Indian," locates studies of the Revolt of 1680 within the larger context of contact period historical and archaeological studies. In describing the theoretical orientations and influences of Borderlands historians and processual archaeologists, I explain how each of these fields has approached the subject of contact period social interaction between Native peoples and Spanish colonists. For different reasons, both historians and archaeologists have avoided critical discussions of the use of violence by Spanish colonists as a means of controlling Native peoples and asserting dominant social and economic relationships. Many of these interpretations tend to refute or minimize colonial violence and deploy "Black Legend" interpretations of contact period, colonial practices.

Rather than critically reading these texts from the vantage of Indigenous peoples, both fields tend to emphasize disease and acculturation as the largely passive, accidental agents of social change. Without a critical examination of colonial violence and conflict, long-term studies of Indigenous resistance have remained unexplored. As a result, events such as the Revolt of 1680 are portrayed as isolated eruptions of anticolonial violence or, alternately, as the fatal spasms of resistance by a doomed cultural system. In this chapter, I explain why contextual studies of the Revolt have never been attempted, how historians have approached Indigenous peoples within Borderlands history, and why archaeological studies which link contemporary Indian peoples to the past have been abandoned by archaeologists since the 1960s.

In the second half of the chapter, I address some the assumptions, limitations, and consequences involved in contemporary contact period archaeology. I discuss the early use of the direct historical approach in southwestern archaeology as well as the emergence and subsequent dominance of the processual approach as a method of interpreting contact period social change. In varying degrees, the processual emphasis on disease, demographic collapse, and acculturation justified a theoretical disengagement with living peoples. Avoiding the subject or study of Puebloan resistance, these scholars dismissed the relevance of contemporary Puebloan traditions as either uninformative or misleading. But like many of the central concepts in anthropology, the concepts of acculturation has been challenged and redefined since the 1960s. In particular, studies of colonial societies (e.g., those that existed in New Mexico in the sixteenth and seventeenth centuries) by social anthropologists suggest that colonized peoples actively resist many colonial cultural influences. As an alternative approach to the interpretive frameworks used by historians and archaeologists, I argue for the centrality of theories of cultural interaction and resistance collectively known as *ethnicity* within Indigenous archaeology.

These issues are further developed in chapter 3, "Explaining the Persistence of Indian Cultures: Ethnicity Theory, Social Distance, and the Myth of Acculturation." Some of the most vexing challenges faced by archaeologists in a post-NAGPRA age involve the complex and dynamic relationships that exist between contemporary Indigenous peoples and the material remains studied by archaeologists. How do we account for changes in these material inventories through time? What social processes facilitate and restrict the exchange and flow of goods and materials across the social boundaries defined by family, kin, clan, and community? How do we account for the stability and reproduction of forms and patterns of material culture without drifting into essentialism? While processual interpretations of material culture emphasize the selective advantages proffered through experimentation, innovation, and exchange, little attention has been devoted to the symbolic associations of artifacts and the role these associations play in the ebb and flow of cultural interactions.

Artifact assemblages analyzed at Old Cochiti clearly reflect a rejection of "technologically superior" European materials, largely because of their location within an alien and oppressive symbolic and social universe. In examining the behaviors of the Pueblos during the Revolt, we are able to situate these behaviors and practices within a historical matrix and move discussions of material use to a more sophisticated level of analysis. Furthermore, in documenting the correlations between the activation of a pan-ethnic movement and the creation of Old Cochiti, we witness the generation and establishment of a social boundaries and boundary maintenance at the community level. Moving beyond traditional, archeological studies of ethnicity, many of which attempt to refine the signals or *diacritics* of ethnic groups, I argue for a contextual study of the processes and archaeological correlates of boundary maintenance and ethnic conflicts.

Read within this context, the subject of colonial violence and conquest narratives acquire a new meaning. Rather than simply enforcing the dominance of one group over another, violent military engagements provide the basis for collective narratives of subordination. These experiences then serve as the developmental matrix for powerful countervailing social movements; these movements appropriate the organizational structures of colonial societies in an attempt to subvert the dominant social order. In this sense, ethnic conflicts are an integral element of human interaction and are generated frequently by the processes of migration, outsider group ascription, conquest, and the subordination of other groups.

Subordinate groups resist these influences with a variety of collective behaviors that can be studied through contextual, historically situated, archaeological studies. These behaviors include the maintenance of social distance through regional abandonments or "refugee movements" to areas beyond the control of the dominant group; alliance formation and the settlement of new communities among subordinate groups; or, as is the case in the Revolt of 1680, the forced relocation of the dominant group through armed conflict. Read in this manner, the abandonments encountered by Spanish colonists and reproduced in historical analyses as evidence of demographic collapse actually reveal a very different set of motives, meanings, and social consequences for the persistence of cultural traditions and Pueblo identities.

In chapter 4, "The Contexts of Subordination and Conquest: The *Reconquista*, Silver, Slavery, and Rebellion in the Spanish Borderlands," I apply theories of ethnicity to the study of colonial encounters and establish the historical context as well as the economic and philosophical justifications for Spanish conquest. One of the more controversial subjects in contact period studies is the degree to which the ideology of Christianity was deployed to support the military conquests of Native peoples. As Spanish colonial scholar Lewis Hanke has demonstrated, the creation of "the Indian" as a subordinate social category was highly contested. In contrast to both "Black" and "White" revisionist historical traditions, some of the most spirited debates regarding the treatment and policies of Indian peoples was carried out by Spanish jurists and legal scholars. The status of Indians was shaped by the terms of social difference or *diacritica* relevant to Spanish society; the "Indian" increasingly came to embody everything the "the Spaniard" was not.

Despite the period of reasoned debate among Spanish jurists and theologians (embodied in the disputation in the Court of Castile between Sepulveda and Las Casas), hierarchies between Christians and non-Christians, between Africans, Indians, and Spaniards, were an integral aspect of European social life. These hierarchies allowed for differential treatment of subject populations and generated the context in which state sanctioned violence was legally and morally justified. In Spain, critical discussions of colonial expansion and the establishment of African slavery among other European powers were overshadowed by the potential for profitable economic ventures. As I assert in this chapter, the use of "official" or "state-sanctioned" violence is characteristic of all nation-states. Without a contextual study of the processes of ethnic subordination, balanced studies of Native resistance movements such as the Revolt are impossible. Later, I explain the importance of silver mining and Indian labor shortages as well as the consequences of forcible conversion and the illegal slave trade in the frontier economy.

In chapter 5, "Abandonment as Social strategy: Colonial Violence and the Pueblo Response," I explain the context in which the initial contacts between Spanish explorers, missionaries, "adventurers," and Pueblos took place. The mining communities of the northern Borderlands served as the staging grounds for early expeditions into the Pueblos region. After examining and documenting the responses of Pueblos to social violence, economic subordination, and forcible conversion, I suggest that the character of these encounters was similar to the pattern of interaction between Spanish adventurers and Native communities in the mining regions of northern Mexico. Perhaps the most informative behavior observed during the entradas was the use of mobility and social segregation as a means of maintaining social distance and preserving the human and cultural resources of the Pueblos. In particular, following a series of violent confrontations in the "Tigua" and Piro provinces, the Spaniards encountered empty villages or communities in which the women and children of the pueblos have been removed.

Following the pattern of ethnic conflict*s* outline in chapter 3, violence and social subordination facilitate social actions and movements that segregate and insulate subordinate ethnic groups. These include large-scale migrations, alliance formation, compartmentalization, and the movements of whole communities to areas beyond the control of the dominant ethnic group. Using primary historical sources, I explore the correlation between social violence and regional abandonments, and I document the economic motives, and social consequences of both the Spanish entradas and colonization of the sixteenth and seventeenth centuries. Building on this argument, I critically examine the contexts in which early population estimates were made and advocate a more cautious, contextually situated application of historical data to archaeological studies of the contact period.

In chapter 6, "Communities in Conflict: Regional Analysis of Revolt Period Settlements," I attempt to answer the questions raised in chapter 5. For example, if mobility is a manifestation of boundary maintenance, where did Puebloan people reside during times of conflict? In documenting regional Revolt period settlements in the Jemez Mountains,

I provide archaeological evidence supporting repeated and strategic occupations of large village communities in the remote mountains surrounding colonized Puebloan communities.

Following a pattern characteristic of these kinds of pan-ethnic movements, the alliances that are established are often fragile. The call for a return to orthodox practices generates different notions of identity and community. As soon as the source of inequality (in this case the Spaniards) is removed, the pan-ethnic movement begins to break apart. After the immediate objectives of the movement are achieved, local village leaders appear to have been unwilling to accede power to the leaders of the rebellion, and local identities are readopted by various pueblos.

In chapter 7, "The Archaeological Correlates of Ethnogenesis: Community Building at Old Cochiti," I demonstrate how the processes of ethnic conflict and ethnogenesis are manifested at the site level. Possibly reflecting the fragility of the pan-Puebloan alliance, the village is fragmented into two communities. The larger settlement contains ritual architecture and was constructed with a high degree of planning and architectural integration. The smaller, secondary settlement has a more irregular arrangement and lacks both plazas and ritual architecture. Equally important, both communities are segregated by significant distances. I suggest that these patterns demonstrate the fragmentary and contested nature of ethnic movements. Equally interesting are the patterns revealed by artifact analysis. Suggesting that Po'pay's call for a return to traditional subsistence practices and the elimination of Spanish cultural influences was widely accepted, the material culture present at each site reflects the complete rejection of Spanish technologies, food, and ritual items.

In chapter 8, "Repatriating Old Cochiti," I argue that the future of archaeology in North America requires an active engagement with a Native peoples. In generating critical analyses of historical, archaeological, and anthropological narratives, Indigenous archaeology moves well beyond the often limited and confrontational dialogues imposed by NAGPRA and highlights the ways in which archaeologists can make valuable contributions to the fields of ethnography, history, colonial history, and postcolonial and Indigenous theory. The impact of this work on these disciplines is also substantive. Using the archaeological, historical, and ethnographic data presented in my study, I suggest that we reevaluate both the importance of violence and the role of Puebloan resistance in explaining postcontact social interactions between colonists and Native peoples. Using this new model of movement and self-segregation, I argue that the traditional troika of Indian destruction (i.e., acculturation, disease, and military conquest) should be reexamined in other colonial contexts. Finally, I demonstrate how this scholarship resulted in the repatriation of the village known as Old Cochiti and discuss the ways in which collaborative projects with Indigenous communities are transforming the discipline.

CHAPTER 2

Creating the Invisible Indian

The impact of Spanish invaders on the people they called "The Pueblos" was nothing short of catastrophic. . . . For over 80 years, we endured until we could endure no longer the threats and acts of religious persecution introduced by the Franciscans and reinforced by the military. Then the unthinkable occurred. Pueblo communities, otherwise very independent of one another, united into a single force to cast out a common foe. . . . Historians tell us that this was the famous Pueblo Revolt of 1680. To my people it was the declaration of independence! That heavy-handed relationship of the Spanish towards the Pueblos was never to be again, even after their return in 1692.

Governor Joseph H. Suina, Cochiti Pueblo (Preucel 2002: 213)

Despite the persistent attention of several generations of historians to the subject of the Pueblo Revolt, archaeological documentation of historic period Indigenous settlements and histories have been slow to materialize.[1] While Hispanic and Latino scholars have made important inroads in the expansion of early colonial voices and perspectives, Native American scholars have yet to assert the kinds of interpretations of contact and colonization offered by Governor Suina. The idea that the past as rendered by historians and archaeologists is informed by contemporary political interests is not new to either field. As historian R. G. Collingwood suggested in a series of critical essays regarding the philosophy of history, the past is always created in a politically and historically situated present. "The past which the historian studies is not a dead past, but a past which in some sense is still living in the present" (Collingwood 1946: 23–24). Anticipating to some extent Foucault's analysis of the production of knowledge (1970, 1980), both Collingwood and historian Edward Hallett Carr's (1961) criticisms of historical scholarship had a transformative effect on the discipline of history. Since the decolonization movements of the 1940s, anthropologists have been engaged in a similar criticism of the representational authority of the ethnographer. Both fields have recognized that the relationships between the present and the past, between subject and object, between "object-makers" and their audiences, is dialectical. Histories are deeply ideological. As state-sponsored projects, as official records of "true" events, and as mythological texts, histories situate groups in relation to one another and affirm the morality and "goodness" of those who write them.

Responding to these criticisms, various postprocessual approaches to the practice of archeology have attempted to reconcile the complex relationships between the past and those who represent it.[2] Unfortunately, much of this critical dialogue has yet to be incorporated by either Borderlands historians or Southwest archaeologists. Both fields can be said to have been informed by political interests that valorize, minimize or ignore the social consequences of colonial expansion (Borderlands histories) or theoretical engagements that, in the name of science, place the pursuit of universal human behaviors above the interests of Indigenous peoples (archaeology).

The efforts of early Borderlands scholars to integrate New Mexico's colonial history with that of the East was thwarted by negative characterizations of Spanish character, colonial objectives and policies. The tensions between the competing colonial narratives of an Anglophile, Protestant East and a Catholic, Hispanic West found expression in several generations of historical texts. As Borderlands historian David Weber argues: "Historians interested in America's Spanish-Mexican heritage have found it difficult to accept the popular reconstructions of the American past that place America's colonial origins entirely within the thirteen colonies, or that omit Mexican Americans from historical narratives" (1987: ix).[3] In an effort to counter the Anglo-centric East-to-West model of American colonial expansion made popular by Fredrick Jackson Turner (1996) and to emphasize the historical and cultural contributions of Hispanic Americans to American social life, Borderlands historians responded with an equally strenuous, revisionist bias.

In the mid-1980s, historian Benjamin Keen (1985: 658) argued that Borderlands histories have been characterized by explicit or implicit attempts to counter negative portrayals of Spanish colonialism and national identity. Building on Keen's thesis in the following discussion,

I argue that these efforts are manifested in three revisionist traditions—each of which deploys a different definition and application of what has come to be known as the "Black Legend," and each of which reflects the political and social investments of Euro-colonial scholars. The idea of the Black Legend is slippery. At times, the legend refers to a well-established tradition of anti-Hispanic sentiment expressed by northern European and American historians. Other historians (Weber 1982) equate the legend with the claims made by Bartolomé de Las Casas (1552) that the excessive violence of Spanish conquistadors resulted in massive population crashes. Still others (Cook 1976, 1978; Borah 1951, 1963; Simpson 1929, 1934, 1952) accept Las Casas's population estimates and argue that European diseases, not Spanish colonial violence, decimated Native populations. In examining the inconsistent and often ambiguous usage of the term, I argue that the term has lost all meaning, except to suggest that Spanish colonial violence was a fiction (a legend) or used to slander those who speak about colonial violence as biased against Hispanics in some way. This curious rhetorical assertion equates criticisms of colonial violence with an attack on Hispanic character. This absurd notion suggests that *mestizos*, peoples of African descent and Indians—those who suffered colonial violence—are somehow not Hispanic. In "legendizing" or fictionalizing violence, historians may avoid being so maligned, but in embracing and promoting disease as an "explanation" of Indian cultural collapse, they are both missing the truth and helping to perpetuate a mythology that is equally damaging to Native Americans: the Invisible Indian. Contact period scholars should instead attempt to disentangle studies of Pueblo population estimates and epidemics and frankly reexamine the use of violence by Spanish colonists.

I begin this chapter with a discussion of the origins and development of Black Legend revisionism in Spanish colonial scholarship and contrast this version with that employed by contemporary North American archaeologists. Representing the first revisionist trend, the works of Bandelier, Lummis, and Bolton alternately celebrated the achievements of "Spanish Pioneers" in celebratory and "triumphalist" conquest narratives and minimized the oppressive and repressive aspects of colonial rule (Keen 1985: 664). The second revisionist trend, characterized by the Berkeley school demographic studies, introduced interdisciplinary methods to the study of contact. Although many of these works suggest that Spanish policies accelerated the destruction of Native communities, the Berkeley school offered evidence for massive, disease-motivated population crashes in the Valley of Mexico. While the methods and interpretations of Berkeley school demographic studies have been attacked by historical demographers (Zambardino 1980; Henige 1986), interpretations of contact interaction that emphasize disease and minimize conflict have largely come to characterize contact period archaeological studies (Creamer 1994). As I argue later, this interpretive framework, reproduced and adopted by processual archaeologists, has thwarted discussions of colonial violence and Pueblo resistance and reiterated a conceptual break between an Indigenous archaeological past and a historical present.

In the second part of this chapter, I argue that the information gap regarding the archaeological study Puebloan resistance and the Revolt of 1680 is the result of the dominance of processual approaches to the archaeology of the Southwest. In an effort to establish archaeology as a nomothetic, scientific study of human behavior, processual archaeologists distanced themselves from the ethnologically, historically based studies of their intellectual ancestors. Historical archaeology and interaction with contemporary Indians were viewed as largely irrelevant to the intellectual project of developing a universalizing story of human adaptation and cultural evolution. The rejection of what has been termed the *direct historical approach* (DHA) to Puebloan prehistory was based on the idea that both disease and acculturation radically transformed Puebloan societies, rendering the use and study of post-contact materials theoretically obsolete.

The use of disease by archaeologists and Borderlands historians to signal the end of an Indigenous past and the beginning of colonial present is informed by both political and professional interests. The positioning of archaeology as a scientific, as opposed to a humanistic, field created new funding, research, and professional opportunities for archaeologists and greatly aided the establishment of a theoretically and methodologically distinct field of study. While the study of "culture contact" has been reinvigorated by the passage of NAGPRA, there exists little enthusiasm among archaeologists for a return to the kinds of ethnological studies that characterized the formative years of the discipline (Goldstein 1998). In addition, there are strong professional disincentives against studies that further establish connections between the archaeological past, conceived of as the data set used by archaeologists, and the microhistories of contemporary Native peoples. The issue is further complicated by the fact that the vast majority North American archaeologists practicing in the Southwest are of Euro-American descent (Blakey 1990; Gero 1983). As NAGPRA has demonstrated, the relationship between the "cultural patrimony" and "sacred objects" of Indigenous peoples and the material cultures studied by archaeologists are complicated by the power relations between colonial and colonized peoples.[4]

Making a Myth Out of Colonial Violence: Invoking the Black Legend

> The statements of closet historians that the Spaniards enslaved the Pueblos, or any other Indians of New Mexico; that they forced them to choose between Christianity and death; that they made them work in the mines and the like—are all entirely untrue. The whole policy of Spain towards the Indians of the New World was one of humanity, justice, education and moral suasion. . . . That we have not given justice to the Spanish Pioneers is simply because we have been misled. They made a record unparalleled; but our text-books have not recognized that fact.
>
> Charles F. Lummis, *The Spanish Pioneers* (1909: 2)

> Only since the invention of the printing press has a successful campaign of defamation lasting centuries been waged against an entire people. That nation is Spain, and the campaign of calumny—known to modern historians as the black legend—made Spaniards pariahs and demeaned the character of Spanish people. This myth, I am convinced, has influenced earlier generations of scholars to cast a cold eye on the achievements of our Spanish Pioneers.
>
> Stewart L. Udall, former secretary of the interior, supervisor of the Bureau of Indian Affairs, *To the Inland Empire* (1987: 204)

References to what has come to be known as the "Black Legend" are a persistent theme in studies of early colonial Spanish history. As the statements of Lummis and Udall, located at opposite ends of the twentieth century, suggest, the issue of the treatment of Native peoples by Spanish conquistadors remains a controversial and politically sensitive subject. Although Lummis's and Udall's statements reflect similar sentiments, the use of the term *Black Legend* did not appear until the 1920s. Since then, the term has come to signify an often confusing, ambiguous, and nebulous array of historical claims and counterclaims. In some cases, the legend refers to criticisms of Spanish Inquisitorial practices leveled by Protestants during the Reformation (Juderias 1917). For others, it refers to Spanish colonial policies toward Indians.

Udall and Lummis define the Black Legend as emanating from a claim that Native peoples suffered violence, slavery, and death at the hands of Spanish colonists. For them, the legend supposedly results from the efforts of Anglo-American and European historians to denigrate the achievements and character of Spaniards and, by proxy, contemporary Hispanic Americans. They suggest that we "celebrate the achievements of Spanish Pioneers" and characterize statements to the contrary as emanating from "racial prejudices" and an Anglophile "tradition of anti-Hispanic sentiment." In *To the Inland Empire: Coronado and Our Spanish Legacy* (1987), Stewart J. Udall argues that the Columbus quincentennial was an occasion "to set aside at last our anti-Hispanic cultural bias, the legacies of centuries of national and colonial rivalry," and reclaim "America's Lost Century." Once again, the defense of colonization is equated with a defense of Hispanic peoples—apparently articulating the negative experiences of Indians and African Hispanics (who are somehow not Hispanic in these statements) is "anti-Hispanic":

> With the Hispanic segment of our population increasing each year, the gains would be substantial if we had the wit to widen our horizons and pluck our Spanish century from the wastebasket. Who, one is emboldened to ask, would be harmed if our national story began with Ponce de León in 1513 instead of with the incoming English at Jamestown almost a century later? (201)[5]

Archaeologists have been equally unclear in their usage of the term. In "Re-Examining the Black Legend: Contact Period Demography in the Rio Grande Valley of New Mexico," Winifred Creamer states, "One of the widespread assumptions of the Black Legend is that Spanish explorers were responsible for the precipitous decline in population that occurred among the indigenous people of the New World *because they spread European diseases to native people who had no resistance to them*" (1994: 263, my emphasis). No mention is made of violence or anti-Hispanic propaganda. For Creamer, the Black Legend seems to refer to the agent, in this case disease, responsible for postcontact population losses. As is often the case with archaeological studies of contact, the experiences of Native peoples as recipients of the often violent processes of subjugation, military conquest, and forcible conversion are marginalized in favor of a more politically neutral and more "scientific" agent of colonial destruction: disease.

Still, the invocation of Black Legend revisionism can best be understood with reference to its original usage. According to historian William Maltby (1971), the term first appeared in an article defending Spanish nationalism. Published by journalist Julían Juderías, *La Leyenda Negra* (1917) defined the legend as emanating from "distortions of historical fact committed by enemies of Spain" that characterized Spain as "the home of ignorance and bigotry, an intellectual wasteland incapable of taking its place as a modern nation" (20). Juderías's work defended the virtues of Spanish nationalism in relation to the Anglophile, Protestant nations of England and the Netherlands, but it made no reference to the conquest of the Americas. For Juderías, the legend was the manifestation of ancient national rivalries between a Catholic Spain, ruled by Phillip II, and an emergent Protestant Europe in the sixteenth century. Many Enlightenment figures targeted the efforts of Phillip II to enforce Catholic orthodoxy through the establishment of the Inquisition as well as the intellectual conservatism of the Catholic Church. Other works, such as Carbia's *Historia de la Leyenda Negra Hispano-Americana* (1943), discussed the development of anti-Hispanicism in Latin America during the Wars of Independence.

Maltby makes a compelling argument that the spread of anti-Spanish propaganda was fueled by the development of new print technologies in northern Europe. As copies of Las Casas's *Breve relación de la destrucción de las Indias occidentales* (1552) were distributed and translated throughout Europe, writers such as John Haklyut (1583) published passages of Las Casas's narrative describing the slaughter of thirty to fifty million Natives in the Americas.[6] While Las Casas's work was written as part of a dialogue with the Council of Indies and Charles II in which hyperbole was an important persuasive tool of discourse, his extremely high population estimates added the dimension of demographic collapse to discussions of the Black Legend. Demographic hyperbole, often supported by the decontextualized mining of high population estimates, remains an essential feature of all disease-based demographic collapse models.

Maltby's work established a separate point of contention regarding the use of the term *Black Legend* in the Southwest. While the Las Casas's descriptions of Spanish violence were deployed by Spain's colonial rivals to undermine the moral authority of the Catholic kings, the source of discussions of violence directed toward Native peoples prior to and following the Revolt of 1680 have nothing to do with Protestant criticisms of Spain or anti-Hispanic sentiment. They are drawn from the internal memos, *relaciones*, personal letters, and historical documents *created by the participants themselves.* These documents were produced in an entirely different social and historical context than Haklyut's inflammatory tracts. Despite this important contextual difference, the legend continues to be deployed in studies of early colonial or contact period histories—effectively short-circuiting discussions of the deployment of colonial violence on Indians or the social consequences of these actions. In this manner, Indigenous rebellions appear as random, spasmodic contractions of a dying culture or, worse, as resulting from the barbaric nature of the savage himself. This juxtaposition of enlightened colonist and savage Indian is part of the mythology known as "triumphalism" and appears with the first wave of Borderlands historians—scholars who resuscitated Spanish colonization at the expense of Indian peoples.

Triumphalist Revisionism and the Birth of the Borderlands: Bandelier, Lummis, and the Boltonian Tradition

> In this country of brave and free men, race prejudice, the most ignorant of all human ignorances must die out. We must respect manhood more than nationality, and admire it for its own sake wherever found . . . and the Spanish Pioneering of the Americas was the largest and longest and most marvelous feat of manhood in all history.
>
> Charles Lummis, *The Spanish Pioneers* (1909: 12)

The words of Charles Lummis, author of numerous studies on the early history of New Mexico and an intellectual disciple of Adolf Bandelier, are easily criticized by contemporary historians. Lummis was the product of a different generation. His celebratory account of Spanish colonialism, aptly entitled *The Spanish Pioneers* (1909), must be considered with reference to the historical and intellectual currents of his day. The influence of social Darwinism and Romanticism, the development of modern nationalism, and the expansion of European imperialism in the late nineteenth century help explain his perspective. Both Keen (1985) and Adelman (1999) argue that like Lummis, early Borderlands studies demonstrate several revisionist tendencies: they tend to assume a priority of Spanish over Indian interests, they are generally unsympathetic to Indian perspectives, and they minimize the oppressive and repressive aspects of Spanish rule. Furthermore, they demonstrate[d] a "relativism that Shunned moral judgments" as well as a "pragmatic" hard-boiled approach that viewed colonial conquest as "unfortunate and inevitable facts of life" (Keen 1985: 664). The justifications for colonial expansion were clearly informed by social Darwinist theories (Darwin 1871), that remained in vogue until the 1940s.

No study of the historiography of the Southwest would be complete without reference to Adolph Bandelier (1840–1914). It was Bandelier who, in collaboration with Lewis Henry Morgan, helped shape the attitudes of American readers on the social complexity of Aztec society in relation to European civilizations.[7] In *Ancient Society* (1877), Morgan located Aztec society on an evolutionary continuum. Characterizing all Native civilizations as forms of barbarism, Morgan suggested that the replacement of Indian societies by more "advanced" European civilizations was part of a "natural" socioevolutionary trajectory.[8] Bandelier was also one of the first American historians to attempt to refute negative characterizations of Spain's colonial empire, suggesting that through the efforts of Spanish colonists, "the Indian emerged fully able to grasp every advantage that civilization of the nineteenth century had to offer" (Bandelier 1885). The work of both Lummis (1956, 1963) and Bandelier foreshadows the "triumphalist" histories of Fredrick Jackson Turner (1996) and his student, Herbert Eugene Bolton.[9]

Bolton was the first historian to apply Turner's classic notion of the frontier to the colonial expansion of New Spain. Furthermore, Bolton established Borderlands studies as a legitimate field of historical analysis. As has been argued in numerous historical syntheses of Bolton's work (Weber 1986; Worcester 1991), Bolton felt that "Turner's east to west model of American development, shortchanged the divergent sources of European expansion" (Adelman 1999: 814). Wary of the implications and controversies surrounding the treatment of Native peoples at the hands of Spanish colonists and eager to elevate the status of Spanish colonial history, both Bolton and John Francis Bannon (1970, 1978) propagated an equally strenuous triumphalist and "Christophillic" revisionist tradition. For Bolton, the entry of Spanish colonists signaled the moral and material salvation of Native peoples.

A sample of Bolton's work helps illustrate the point. In *The colonization of North America, 1492–1783* (1919), Bolton stated that "the [Spanish] kings desired to convert, civilize and exploit the natives, but it was soon found that if the savage were to be converted, or disciplined, or exploited, he must be put under control. . . . If the Indian were to become either a worthy Christian or a desirable subject, he must be disciplined in the rudiments of civilized life" (44). Later Bolton defended Spain's colonial policy adopting a "white man's burden" defense of Western colonization:

> Hr colonial policy, equaled in humanitarian principles by that of no other country, perhaps, looked to the preservation of the natives and to their elevation to at least a limited citizenship. . . . Such an ideal called not only for the subjugation and control of the natives, but for their civilization as well. . . . The essence of the mission was the *discipline* religious, moral, social and industrial which it afforded. (53)

According to Bolton, Spanish missionaries allowed Native peoples to transcend an aberrant moral state, introduced agriculture to Native peoples and helped civilize the continent. Missionaries "explored the frontiers, promoted their occupation, defended them and their interior settlements, taught the Indians the Spanish language, and disciplined them in good manners, in the rudiments of European crafts, of agriculture and even of self government. . . . Moreover, the missions were a force which made for the preservation of the Indians, as opposed to their destruction, so characteristic of the Anglo-American frontier. . . . They were a conspicuous feature of Spain's frontiering genius" (61). Unfortunately, in attempting to "Parkmanize" and elevate narratives of Spanish colonization, Bolton's studies ignored the contributions and perspectives of mestizos, Native peoples, and women (Hurtado 1993, 1995).[10]

Bolton's legacy is difficult to overestimate. Among the 427 graduate students he trained while at Berkeley are some of the most prominent figures in Borderlands and Revolt period research. Charles Hackett, G. P. Hammond, J. M. Espinosa, W. W. Borah, and John Francis Bannon were all students of Bolton (Bannon 1978: 278–290). After Bolton's death in 1953, Bannon (a Jesuit priest) continued Bolton's work. As a staunch defender of Spain's conquest and missionary practices, Bannon actively resisted critical interpretations of Spanish colonial activities (Weber 1986, 1988; Bannon 1978).[11]

Narrating Demographic Collapse: The Berkeley School and Disease as the Agent of Destruction

Bolton's legacy extends to the second wave of revisionism. Along with A. L. Kroeber and geographer Carl Sauer, Bolton founded the journal *Ibero-Americana*. Unlike Bolton's more conservative Borderlands works, these Berkeley school demographic studies incorporated an interdisciplinary approach to contact period social change. Signaling a break with the Boltonians and triumphalism, these studies at times offered veiled criticisms of Spain's colonial policies but, more importantly, offered evidence supporting massive disease-motivated population crashes in the Valley of Mexico in the sixteenth century (Sauer 1935, 1948; Cook and Simpson 1948; Simpson 1952; Borah and Cook 1963).

Despite the fact that the methods, statistical projections, and conclusions of the Berkeley studies have been widely criticized by historical demographers (Henige 1978, 1986: 700–721; Farris 1978: 217–237; Zambardino 1980: 1–27), the disease-motivated population crash model has been widely adopted by archaeologists and historians.[12] The notion that diseases wiped out all or most of the Indians in the Americas is a belief widely held by most people. The substitution of disease for violence (and stories of Indigenous resistance) has both supported the notion that Indians are no longer with us (4.5 million in the United States) and that postcontact Indian histories are either nonexistent or irrelevant. These are absolutely pervasive themes associated with Indian peoples in textbooks, film, and other media and they contribute directly to the mythology of the invisible Indian. Because much of this scholarship originated in the academy, it has acquired an unquestionable, gospel-like character in academic circles. And it is even repeated by Indian scholars themselves (Deloria 1969).

The trend toward disease as an explanation of postcontact Indian cultural collapse became popular in the 1960s, 1970s, and 1980s. As the New Archeology sought explanations of culture change that could be applied universally to human societies, local histories and conversations with descendant communities were abandoned. The explanation of demographic collapse carried with it the assumption that Indian people had somehow "lost their culture" through either acculturation or population decline. Curiously, change, a universal element of any human society, is always depicted as a *loss* for Indian people. This essentialist reading of Indian culture always attempts to locate Indian authenticity in the past—a contested landscape in which archaeologists are able to assert a dominant voice. This tension is at the heart of many Native peoples' discomfort with archaeological studies. For some Indigenous peoples, it represents a colonization of ancestry (see McNiven and Russell 2005). Studies of disease also allowed archaeologists to extend their analysis of "the environment and adaptation to microscopic levels"—the language of disease is the language of science, and it helped bolster the authoritative voice of a field too often associated with the collection and display of beautiful things and alien treasures.

Archaeology has always struggled to assert an identity distinct from amateur collection, the assembly of rare geographically oriented specimens had led to simplistic (or nonexistent) explanations of culture change. Earlier generations of scholars (Taylor 1948; Ford 1952) felt that the preoccupation with chronology and stylistic typology had eclipsed more meaningful and interesting anthropological objectives. These criticisms led to the development of new methodologies and research designs devoted to the explanation of culture process (Binford 1964: 425–441). These insights resonated with a new generation of scholars trained in the 1960s, and the tenets of the "New Archaeology" became immensely popular in American academia (Redman 1991: 295–307). These scholars redefined the objectives of southwestern archaeology to include a more "scientific" approach to data collection and interpretation grounded theoretically in logical positivism (Preucel 1993) and distinguished from the classificatory-descriptive and culture-historical frameworks of previous generations of scholars (Willey and Sabloff: 229). Moreover, these scholars felt that the search for universal process should be grounded in anthropological theory and that the explication of stylistic variation and the typological distributions of artifact styles assumed an "aquatic view of culture" in which the mechanisms of culture change were entirely unexplained (Binford 1962: 217–225).

Borrowing from the evolutionary framework of Leslie White (1959) and the ecosystems models of Julian Steward (1955), these "processual approaches" have been widely applied by American archaeologists since the late 1960s. These interpretations view prehistoric social interaction and settlement shifts as elements of culture process. In these reconstructions, environmental stresses, technological adaptations, and population pressure interact in varying degrees to determine human adaptation (Cordell 1984). The interplay of these variables moves prehistoric settlements along an ecosystemic continuum away from or toward a conceptual state of equilibrium. Within this interpretive framework, changes in settlement patterns and periods of aggregation and abandonment are interpreted as adaptational responses to climactic shifts. Many of these interpretations assume the maximization of economic or technological efficiency or advantage within a "marginal" environment as determining the patterns and processes of prehistoric social integration an intracommunal interaction. These models cast

Puebloan peoples as competitive entities who struggle with each other and their environment for a share of limited resources—sometimes to fantastic and disastrous consequences (Wicox 1994).

Mandated by the passage of NAGPRA, the application of contextual or interdisciplinary methods to the study of historic period, social interactions between Native peoples and colonists has created both methodological challenges and new opportunities for archaeologists. Among the most difficult issues have been the assessment and integration of widely ranging historical, ethnographic, and archaeological materials. For archaeologists working in a post-NAGPRA age, the assessment of multiple lines of evidence in determining issues such as "cultural affiliation" has become an increasingly important and controversial area of archaeological research. One of the problems faced by archaeologists and museum workers has been the development of appropriate methods and models of contact period archaeological analysis. Despite the mandates of the legislation, there are strong disincentives for archaeologists to conduct studies that document clear associations between the Puebloan prehistory and contemporary societies. As Lynn Goldstein stated in a discussion of archeology in the post-NAGPRA age:

> The archaeology of the 1980's and 1990's not focused on linking archaeological cultures to modern-day descendents. . . . In fact, modern archaeologists have shied away from making such associations. The associations made by archaeologists earlier in this century are generally seen as naïve and simplistic, and although no one has systematically deconstructed these early works . . . scholars have realized that the job of correcting them would take great energy and diligence. (Goldstein 1998: 373)[13]

The New Archaeology and the End of Indian Histories

The application of the scientific method to archaeological research both increased the opportunities for scientific funding and redefined the relationship between ethnographic information and archaeological research. There were also methodological challenges raised by the interpretation of "static" artifact assemblages using the deductive method. The lack of "observable" patterns of behavior led to the creation of behavioral archaeology (Schiffer 1976) and the use of general and dehistoricized ethnographic analogies to discover patterns and processes of human behavior. Here, the professional and theoretical divergence of social anthropology and the New Archaeology played a particularly prominent role in the disassociation of Indigenous material culture and contemporary Native Americans. Processual research rejected "historical" explication as both narrow in focus and theoretically ungrounded (Binford 1967: 234–235). Once again, acculturation and disease had rendered the Indian invisible and irrelevant.

The use of ethnohistorical evidence and the application of the direct historical approach, or DHA, were rejected as both uninformative and potentially misleading (Binford 1968b). In an effort to expand the informative potential of southwestern prehistory, processual archaeologists asserted that Puebloan people had become too Westernized to contribute to archaeological interpretations. Furthermore, they argued that historical analyses of Indigenous material cultures contributed little to the illumination of universal laws of human behavior.

With a few exceptions (Hill 1970; Thomas 1989; McGuire 1997; Preucel 2001), the informative potential of ethnographic information and the use of the DHA in Puebloan prehistoric reconstructions has been largely abandoned since the late 1960s. In a position paper reflecting this trend in research, Steadman Upham (1989) attacked the application of the DHA to Puebloan prehistory. Citing the demographic reconstructions of Dobyns (1966, 1983) and Ramenofsky (1982), Upham rejected ethnographic characterizations of Puebloan peoples as egalitarian societies in prehistoric times (Bunzel 1938: 352; Dozier 1970a: 209). While contemporary Puebloan people may still exhibit "traditional behavior," Upham stated that these groups "only reflect ancestral adaptations that are now used by descendants in different ways. . . . In this respect, the present is a mirror of the past only insofar as the present can be understood as a dismantled version of the past" (272).

While this position affirms the exclusion of "tainted" ethnographic and archaeological data, this characterization of Puebloan peoples also assumes an essentialist vision of Native peoples as passive recipients of Western culture and makes questionable assumptions about the processes, motivations, and character of acculturation. In a statement reflecting the general interpretive climate of American archaeology, Upham concluded, "The magnitude of change that occurred during the contact period generally precludes the use of Southwestern Ethnographic information in archaeological research" (265). But since the late 1960s, anthropologists have critically reexamined the concept of acculturation, revealing a much more dynamic and negotiative process of cultural interaction. This research has questioned the assumptions relating to acculturation by focusing on the boundaries that alternately separate, distinguish, and unify cultures. The formation and maintenance of these boundaries have been defined by social anthropologists as *ethnicity*.

The processes of ethnogenesis and the maintenance of cultural distinction through ethnic boundaries is of crucial importance in the development of Indigenous archaeology. The exchange of resources and ideas through the networks of religious societies that crosscut Puebloan society was documented in some of the pioneering ethnographies of the last century. Unfortunately, the dogmatic rejection of historical explication (Binford 1968b) and the predominance of ecological modeling have completely excluded the informative potential of contact period archaeology and perpetuated a simplistic portrayal of acculturation and cultural change among Puebloan peoples. The cultural-ecological models so popular in the Southwest today tend to favor those interpretations that deploy economic motives in the explanation of settlement shifts, periods of aggregation, and abandonment without reference to intrasocial processes.

Furthermore, adaptationalist and selectionist interpretations assume an economic rationality for the adoption and spread of technology and ideas without referencing those social and religious institutions that resist social changes While the importance of human and environmental interaction in prehistory cannot be dismissed, these models do little to account for strategies of resistance or the location and function of power relations between colonizers and the colonized. Critical studies of colonization, resistance, nationalism, and ethnicity offer several alternative theoretical approaches to archeologists interested in contact period studies. Unfortunately, studies of ethnicity and boundary maintenance have been only narrowly applied in contemporary archaeological studies of Native American prehistory. Most of the theoretical work dealing with ethnicity and archeology has taken place within the context of European archaeology.

Acculturation and Culture Contact: Archaeologies of Ethnicity

Many of the earliest archaeological investigations in Europe attempted to explore the relationships between archaeological cultures and historically documented ethnic groups. Undertaken in the late nineteenth century, these studies were heavily influenced by the romanticism and nationalism of colonial Europe and have since been criticized as thinly veiled and politically motivated forms of propaganda (Gellner 1987; Smith 1986: 174).

More recently, there has been a resurgence of interest in the relationships among ethnicity, nationalism, and archaeology among European archaeologists (Shennan 1987: 1–40). Sian Jones's (1997) study of Roman colonization in England explores the relationship between changes in material culture and the persistence and reformulation of ethnic identities. North American archaeologists such as Cordell (1991) and Rogers (1993) have begun to reexamine the relationships between historic Native communities and archaeological cultures in more contextual studies of postcontact social interaction between Native peoples and Europeans. Still, the vast majority of contact period archaeological studies are oriented toward the analysis of disease and acculturation as catalysts of cultural change and discontinuity.

Generally, archaeological approaches to ethnicity have focused on the study of stylistic or technological markers of ethnicity as signals or diacritica of specific ethnic groups. This approach attempts to refine our understanding of the diacritica of material culture, as either conscious or unconscious representations of ethnic identities (Wiessner 1983, 1984, 1985, 1990; Lechtman 1977, 1979). Many of the problems associated with this approach revolve around the same issues faced by social anthropologists in the identification of cultural units or ethnic groups. Diacritica are often difficult to single out without significant ethnographic or historical documentation (Chilton 1996). Even with contextual information, archaeologists are still required to write and conceive of ethnic groups (via archaeological cultures) without the benefit of informants' own conceptions of social difference, without direct knowledge of their notions of insider and outsider relationships or, more importantly, whether the differences observed in material culture serve as proxies for ethnic identities.[14]

More recently, archaeologists have attempted to resolve these issues. The most promising (and complicated) application of theories of ethnicity to archaeological studies lies in the application of Bourdieu's notion of *habitus* and theories of practice (1977) in an effort to substantiate the degree to which routinized, culturally specific practices are revealed in technological and stylistic choices (see Bentley 1987: 24–55; Chilton 1998; Gosselain 1992; and Sackett 1986: 268–269, 1990: 33, 37). Leroi-Gourhan's notion of *chaînes opératoires* (1993: 305–319) reflects a similar attempt to articulate a theoretical explanation for normative behaviors.

But as Stark has argued (1998: 1–11), the New Archaeology generally rejected the equation of "distributional patterning with ethnic (or ethnolinguistic) groups in the past" (3; see also Cordell and Yannie 1991; Shennan 1989a, 1989b). As a result, archaeologists have operated without a clearly established theoretical explanation of normative behaviors. The assumption seems to be that the maintenance of different cultural practices are a function of geographic isolation or a failure to adapt to, generate, or adopt more efficient technologies.

In North America, the professional dichotomization of historic and prehistoric archaeology has had a dramatic effect on the scholarship and representation of contact period archaeology. Prehistoric archaeologists traditionally have focused on Native American sites and have generally ignored contact period sites as either uninformative or the domain of historical archaeologists (Lightfoot 1995:199). Historical archaeology in North America has traditionally focused on Euro-American sites. The result is an information gap regarding contact period Native settlements and an artificially constructed dichotomization between historic-prehistoric and European-Native archaeological cultures. This oppositional relationship forces us to conceive of archaeological cultures as monolithic entities—conceptually distinct, but generally uninformed by the intersocietal social dynamics that characterize the contact period.

Among contact period Puebloan sites, this information gap is highly pronounced. While archaeologists in Mesoamerica have incorporated multiple lines of evidence in the interpretation of prehistoric social dynamics, southwestern archaeologists have generated interpretations that reinforce this dichotomy. Lightfoot (1995: 199–217; Lightfoot and Martinez 1995: 471–492) has revived the debate regarding the application of the DHA to multiethnic archaeological sites. Working in a contact period site on the California coast, Lightfoot has stated that, "representing an interface of common concern, culture contact studies may revitalize holistic anthropological approaches that consider multiple lines of evidence from ethnohistorical accounts, ethnographic observations linguistic data, native oral traditions, archaeological materials and biological remains" (1995: 199). Lightfoot characterizes the rejection of the DHA as focusing on "simple," "direct historical," "specific analogy," or "specific historical" analogues (Wylie 1982, 1988: 142, 1989: 10–17). While acknowledging the limitations of ethnohistorical documents, Lightfoot suggests that archaeologists should read critically and "define:

1. the time of the observation,
2. the cultural context in which the text was written,
3. the nature of the explorers journal, administrator's letter, ethnographic report,
4. the training of the observer (explorer, missionary, ethnographer, etc.),
5. the method of observation (participant observation, interviewing elders, oral tradition, etc.) and
6. the degree to which different observations corroborate with one another. (205)

Clearly, one of the most important methodological changes associated with an Indigenous archaeology involves the use of multiple lines of primary source materials as well as an engagement with the theoretical trends and movements within the fields of colonial history, Borderlands studies, and archaeology. A separate yet equally important facet in the recovery of indigenous histories involves a similar critical engagement with the field most commonly associated with Indigenous peoples: anthropology. If historians helped render an invisible Indian, anthropologists have had an equally negative effect on Indians with the creation of the perpetually vanishing Indian.

The often essentialized renderings of Indigenous peoples, a mainstay of North American anthropology since its inception in the 1800s, may have helped generate work for anthropologists interested in the "salvage of the Savage," but they also positioned Native Americans as characters in a perpetual dialectical opposition to modernity. The pervasiveness of the perpetually vanishing, essentialized Indian within the American narrative is itself illustrative of the nostalgia that Westerners feel for the comforts of a less complicated, more noble ancestral imaginary, but it has little to do with Indians or their histories. In fact, a significant portion of the objections voiced by archaeologists to the involvement of modern Indians in archaeology is rooted in the notion that archaeologists have a more intimate knowledge of and connection to the ancestors of Indians than the Indians themselves. For many archaeologists, the *real* Indians, the ones who understand the materials archaeologists interact with, are long gone.

But just as NAGPRA forced archaeologists to confront the very real and contemporary presence of Indians in the late twentieth century, postcolonial (after 1940) anthropology has wrestled with the legacy and meaning of primitivity as a colonial appellation. The presence of contemporary Indigenous communities in a postmodern, postcolonial (and soon to be postglobalization) world demands both explanation and a theoretical engagement by social scientists with that presence. If archaeology is to have any kind of substantive engagement with contemporary Indians, we need to closely examine the role archaeological discourses have played in the construction of *authentic* and *inauthentic* Indian cultures.

If "writing Indians back into the present" is the goal of Indigenous archaeology, we must question the centrality and pervasiveness of acculturation theories within our field. Much of the critical dialogue surrounding the persistence of cultural distinctions in colonial situations was generated by the work of American sociologists and urban anthropologists. Helping to generate a subject of inquiry known as ethnicity theory, these studies transcended discussions of acculturation and the perpetually vanishing primitive by documenting the ways in which humans actively maintain cultural distinctions despite long histories of contact with other groups. If the essentialized Indian is only visible through the prism of acculturation, then an Indigenous approach to archaeology must confront both the origins and implications of these ideas. Ethnicity theory is one approach to cultural interaction that explains how human beings actively and consciously maintain distinct identities. It is the antidote to acculturation.

CHAPTER 3

Explaining the Persistence of Indian Cultures

Ethnicity Theory, Social Distance, and the Myth of Acculturation

If alienation is a malfunction of modern society, ethnicity is an antidote.

Ronald Cohen, "Ethnicity: Problem and Focus in Anthropology" (1978: 401)

The mythology of the "perpetually vanishing primitive" and the idea that Indian peoples are doomed to eventual extinction are pervasive in academic discourse and popular culture. *National Geographic* in particular has become the conduit through which anthropological theories of acculturation and critiques of industrialization and "modernity" have been channeled to an eager and concerned citizenry. Indians almost universally are associated with nature, innocence, and vulnerability; whereas Europeans and Westerners are depicted as agents of modernity, industrialization, and change and are often responsible for the degradation and exploitation of marginalized Indigenous peoples. This narrative is certainly partially accurate. Indigenous peoples frequently have no representation in democratic political systems. Threats to sovereignty and resources are very real. If concern among the public translates into greater visibility for the struggles faced by Indigenous peoples—a position many Indian peoples willingly deploy—then there is seemingly no harm in allowing the trope of the vanishing primitive to remain unchallenged. It is a stereotype that appeals to the sensibilities of the dominant culture and exerts a powerful moral force in shaping public sentiments regarding the plights of Indian peoples.

What is perhaps in need of criticism is the idea that Native peoples have no means of dealing with changes brought about by five hundred years of interaction with Westerners. If Indigenous peoples have been "vanishing" for five hundred years—an irony rarely recognized—maybe the conceptual lens through which Indian cultures are viewed needs reevaluation. Native peoples, like all human groups, do in fact possess the means of regulating interaction with outside groups. *All* cultures do: without these mechanisms, humanity would be culturally monolithic and undifferentiated. It is the maintenance of cultural distinctions and explanations of those differences that have been overwhelmed in the discourses of acculturation and the mythology of the vanishing primitive. In the twenty-first century, one can easily recognize the degree to which contact with Western culture has motivated powerful countervailing movements, often referred to as *fundamentalisms*, which actively mobilize resources—symbolic, ideological, and material—against the influences of Westerners. This notion has never been applied effectively in any popular narrative of Native American survival. Theories of ethnicity and "boundary Maintenance" provide an important antidote to acculturation and the mythologies of conquest.

The concept of ethnicity gained prominence within anthropological literature in the late 1960s. Since that time, the word has filtered into both academia and popular discourse; its shifting, sometimes transient meaning has been used to describe immigrant communities, tribal peoples, fragmented social factions, or, more generally (as is the case in archaeology), as an equally slippery synonym for *culture*. Despite these ambiguities, ethnicity can be best understood as emerging from within a specific historical and theoretical context within the social sciences, largely as a reaction to the limits of structural-functionalism (Banks 1996: 11). Greatly simplified, structural functionalism imagines societies as being loosely analogous to an organism—a holistic and integrated system, internally differentiated by specialized structures and functions—that struggles to maintain equilibrium. As the "grand theory," adopted by American sociologists and British social anthropologists and used to explain both primitive and modern societies, its limitations were more clearly revealed. Functional models were both intricate and logical ("good to think") but also failed to account for change of any kind.

Theories of boundary maintenance, ethnogenesis, and ethnic conflict are crucial elements in interpreting contact period studies in general and the Revolt of 1680 in particular. Defined as a form of social organization, ethnic groups demonstrate organizational capacities similar to other more commonly defined social units such as bands, tribes, and states. More importantly, the collective behaviors and social movements generated by ethnic or pan-ethnic groups demonstrate social processes and patterns of behavior (migrations, resettlements, and communal segregations) that can be interpreted using archaeological methods.

In contrast to their Boasian counterparts, the comparative sociology of British anthropology rendered history, the role of the individual, and "culture" as largely epiphenomenal. In the search for Western societies' primitive social antecedents, internal and external conflict and the processes of social change were treated as largely irrelevant, maladaptive, temporary illnesses of the system. The organic analogy of societies as isolated wholes or as atavistic cultural isolates implied a high degree of stability and social isolation for human communities. The more isolated the primitives were, the more valuable they appeared in the eyes of the ethnographer. As a result, prior to the 1940s, the study of the interactions between cultures remained poorly understood.

But the struggle between the forces of history and culture (via Weber) and structural-functionalism (via Durkheim) were revealed far from Oxbridge, in the turbulent mining towns of Rhodesia. As anthropologists at the Rhodes-Livingstone Institute struggled to explain pan-tribal interactions within colonial territories using the dominant paradigm, they turned to the models of American sociology and the urban ethnography, conflict studies, and "problem-oriented" research of their Chicago school colleagues.[1]

The emergence of ethnicity as a concept can be traced to attempts by American sociologists to understand the social relationships between *ethnic* communities in urban centers and attempts by British social anthropologists to describe *tribal* interactions within colonial territories.[2] While the focus of these studies as well as the intellectual communities in which they developed are diverse, each has contributed an important element to the study of ethnicity. Each shares a common interest in explaining the dynamics of human interaction within and, more importantly, between communities. This early work set the stage for Frederick Barth's (1969) synthetic treatment of ethnicity as a social process. The final part of the chapter takes the work of Barth a step further in the examination of ethnic conflicts and transforms a concept carried "in the head" to a form of identity capable of motivating powerful social movements.

Toward a Sociology of Culture: The Chicago School and the Rhodes-Livingstone Institute

In contrast to British functionalism, which emphasized the study of social structures in "primitive" societies, the work of George Herbert Mead, Robert Merton, Talcott Parsons, Ernest Burgess, and Robert E. Park was more heavily invested in the role and agency of the individual within a modern societies. Influenced by the writings of Freud and the sociology of Georg Simmel, these studies moved beyond the boundaries of functionalism and suggested that social meaning could be illuminated through an examination of the processes of social interaction.[3] In contrast to the functional primacy of social structure, these approaches advocated the idea that free-thinking individuals actively create society through interaction with others. Their work emphasized a dynamic interplay between the individual and the group within a specific cultural and historical context (the largely unacknowledged antecedents of agency and structure after Giddens and habitus after Bourdieu).[4]

In contrast to the static model of functional equilibrium that characterized continental sociology, for the Chicago school theorists, social interaction was interpreted as a fluid process of interpretation, negotiation, and adjustment (Lal 1986: 283). According to Mead's thesis of *social behaviorism* (1934: 260–328), the individual possessed multiple social roles or identities that invested one in a multitude of social spheres. In a later essay, Park suggested that these identities provided the context in which status was affirmed: "an individual may have many "selves" according to the groups to which he belongs and the extent to which each of these groups is isolated from the others. . . . The individual is influenced in differing degrees by the different types of groups of which he is a member" (1921: 102). Most importantly, these facets of individual identity may be antagonistic. Group membership implied both solidarity and a personal investment in a historical and cultural context. "A group may be said to have a life when it has a history. . . . The heritage of every permanent social group is accumulated and transmitted from one generation to another" (Park 1921: 177).

The relevance of cultural and historical context, conflict, and collective action was generated by an interest in solving the social problems posed by the interaction of African Americans and European immigrants in Chicago at the close of the nineteenth century. According to Park, democratic ideals such as collective action and consensus were threatened by the persistence of ethnic cultures. His main concern was to understand (and develop policies to overcome) the physical and social isolation of these groups. Following this theme, the Chicago studies described a process by which immigrant groups formed and maintained stable and persistent communities in a multicultural context through a system of "social boundaries." Using surveys and questionnaires, they correlated stereotypes and patterns of interaction using a social distance scale (Park 1950, 1955). Their data showed that despite the interactions of ethnic minorities in an urban setting, Jews, Italians, Poles, and other groups had maintained cohesive communities and exclusive identities through several generations. Exchanges between ethnic groups seemed to be mediated and regulated by communally defined standards of behavior that insulated the members from outside influences.

These concepts were of great interest to British anthropologists studying pan-tribal movements and conflicts in the mining communities of East Africa's Copper Belt. Here, the colonial context and "tribal" (the non-Western equivalent of ethnic) conflict were pervasive themes of social interaction. Park's social distance scale was adopted by the students of Max Gluckman at the Rhodes-Livingston Institute (RLI) in modern Zambia (Banks 1996: 24; Werbner 1984: 157–185). Clyde Mitchell, Abner Cohen, and A. L. Epstein (the future Manchester school theorists) examined the relationship between colonial governments and indigenous African communities (Kuper 1993: 135–158). Using models and methods borrowed from their Chicago counterparts, their interdisciplinary, field-based research developed long-term analyses of Indigenous peoples in urban communities. Physically and theoretically distant from the traditional structural-functionalist work of Oxford and Cambridge, these Copper Belt studies helped illuminate the dynamic and fluid social processes that Barth would later identify as "ethnicity" in social groups and boundaries.

Gluckman's (1940, 1958) emphasis on the social boundaries between cultures was further developed in Mitchell's (1956) discussion of the signals (diacritica) of social distance. Later, Cohen's (1969) work among Hausa traders echoed the Chicago studies' emphasis on individual agency in shaping the dynamic character of ethnic identities. These three concepts—social boundaries, diacritica, and fluidity via personal agency—provide the basic elements of contemporary discussions ethnic theory and help explain the persistence of Indigenous identities after several centuries of colonization.

History, Culture, and Conflict: The RLI and the Copper Belt Studies

Max Gluckman's classic study of Zulu-European interaction (1958: 12–21) acknowledged both the colonial context in which Zulu-white interactions took place as well as the social boundaries that structured interaction between the two groups. A common criticism of structural-functionalist studies was the notion that these studies selectively emphasized a parochial view of societies as isolated wholes (Malinowski's apt illustration) but ignored the larger social context in which different peoples related to one another. This essentialist portrayal of Indigenous peoples reiterated a primitive/static, European/dynamic false dichotomy. But these early monographs of the RLI (1940) were the first ethnographies to explicitly include the presence of European and colonial peoples as partners in a larger sphere of social interaction. Here the connection with Park's sociological studies becomes clearer. As members of a colonial policy institute, Gluckman and his students were most interested in monitoring pan-tribal movements and explaining the persistence of smaller intertribal factions in large, urban communities.

Building on the work of Gluckman, Mitchell's 1953 study of pan-tribal social movements in the Africa's Copper Belt (*The Kalela Dance*) demonstrated the tenacity and resilience of tribal identities as well as the complex nature in which social identity was communicated. Mitchell's analysis of the dance showed that tribal identities were performed and signified in public displays (the dance is an example of this) in order to affirm membership in a group. In multitribal communities, behavior, dress, and speech were used as signals that identified people with their homeland.[5]

Still, there were some unresolved issues concerning the dynamic nature of ethnic identities. Abner Cohen's *Custom and Politics in Urban Africa* (1969) added the dimension of instrumentality to ethnicity and further problematized previous notions of tribal identity and information flows. His monograph on Hausa traders in the Yoruba-dominated city of Ibadan demonstrated how a pan-Hausa identity could be assumed by non-Hausa immigrant groups (1969).[6] Becoming Hausa allowed a degree of social and economic mobility that was unavailable to the members of smaller tribes. Colonial authorities had facilitated this process in choosing to recognize the Hausa as an "official" tribal group. The political power and authority of these tribal groups was granted and given sanction through the act of recognition (an issue of great importance in the Spanish colonies). When these "inauthentic Hausa" returned to their villages, they completely shed their pan-Hausa identity. In a local context, "being Hausa" and asserting a pan-tribal identity had no meaning.

The differences between these sociological studies and the neatly bounded ethnographies of unacculturated "tribal" peoples were twofold. First, in traditional ethnographies, the anthropologists created and defined the units of study not the informants. The untidy internal divisions between "ethnographic communities" had until now been glossed over or minimalized. Second, ethnographic studies could not assess long-term changes through time; the differential analysis of cultural traits using comparative methods explained little about how change occurred or was resisted. The work at the RLI questioned the processes of social change and forced new debates regarding cultural interaction, conflict, and acculturation. If "ethnic groups" in urban settings were able to change their surroundings and the inventories of their material culture and still function as discrete groups, then anthropologists needed to explain the persistence of cultural differences between cultures. The focus of study should not be on the comparison of cultural inventories but on the processes by which groups defined and segregated themselves. In this context, Fredrick Barth's influential essay, begun as an attempt to clarify comparative methods, helped generate important insights into the processes of information flow and challenged dominant essentialist renderings of Indigenous peoples.

Acculturation and Change: Cultural Units and the Problems of the Comparative Approach

The comparative study of cultural traits both demanded and reiterated a conceptual dichotomy between acculturated Western peoples and the atavistic, primitive, and isolated tribal units of colonial society. As the RLI studies demonstrated, the parochialism inherent in early ethnographic studies created the impression that non-Western peoples were somehow frozen in a primordial state. The focus of anthropology after World War II on savage ethnography was based on the assumption that as the material contents of primitive cultures were transformed through contact with the more developed world, the cultural differences that served as the basis for comparative studies would be erased. To evaluate acculturation, empirically defined attributes, which could be measured and categorized by anthropologists, needed to be standardized. Without this clarification, statistical data were clearly suspect.

Methodological problems associated with the comparative approach—specifically, the definition and statistical comparison of cultural units (Galton's problem)—raised questions about the criteria ethnographers used in the assignment of cultural categories.[7] In addition, the establishment of categories themselves implied a degree of social cohesion, homogeneity, and stability frequently at odds with or completely alien to the categories ethnographic subjects used to define themselves.

Raoul Narroll, well known for using statistical models in the reconstruction of prehistoric populations, directly addressed these challenges in an article published in *Current Anthropology* in the mid-1960s. Here, Narroll was confronted with a common predicament of archaeologists—sampling strategy:

> First, probability sampling theory requires that each sampling unit in the universe sampled have a mathematically definable chance of being selected. Obviously, the first step in meeting this requirement must be to define the units being considered. For cross-cultural surveys this task has never been rigorously accomplished. Secondly, the validity of the results of cross-cultural surveys has been questioned in discussions with anthropologists on the ground that the units were not comparable. (1964: 283)

Other anthropologists had struggled with the issue of cultural units, but their definitional criteria were far from uniform. Evans-Pritchard (1940: 278–279) had described the tribe as a political unit within the Nuer. Some scholars defined cultural boundaries as linguistic units; others were defined by economic self-sufficiency. Reichard (1938: 413f.) defined a "tribe" as a closed society, with laws and morals applying only to its members. Generally, the professional judgment of the anthropologist seemed to determine what the salient features of culture-bearing units were, but each resisted universal definition. In addition, the categories themselves implied a degree of social cohesion and homogeneity frequently at odds with or completely alien to the categories ethnographic subjects used to define themselves. Furnival's comparative study of plural societies in Burma (1948) and Leach's classic Highland Burma study (1954) had raised an issue few ethnographers cared to deal with.[8] The questions raised by Narroll focused the attention of anthropologists on the "edges" that separate and distinguish cultural systems and challenged the prevailing prism through which anthropologists identified Indigenous peoples—acculturation.

Ethnic Groups and Boundaries

In 1969, a small group of Scandinavian anthropologists organized a symposium on this very subject. The introductory essay (Barth 1969) dramatically improved our understanding of ethnicity as a concept and applied a more sophisticated means of addressing the issue of cultural change within non-Western societies. While the connections between Barth and the Manchester school are not completely clear, according to Barth, anthropologists had described cultures as bounded and discrete aggregates of people; these empirically evident categories were heuristically valuable but implied a preconceived view of what the significant factors of cultural identification were. The simplistic classification of "Western" and "tribal" material inventories explained little about the social processes that segregated different groups and ignored Indigenous perspectives on the social significance of these materials. Furthermore, acculturation models suggested that cultural differences developed as the result of geographic or linguistic isolation, but they paid no attention to the social processes that constrain and facilitate change through time. The use of this "taxonomic approach" "limit[s] the range of factors used to explain cultural diversity. We are led to imagine each group as developing its cultural and social form in relative isolation, mainly in response to local ecological factors, through a history of adaptation by invention and selective borrowing" (Barth 1969: 33).

Barth turned the culture concept inside out and questioned the same processes of cultural diffusion (the aquatic view of culture) that Binford had attacked in the New Archaeology (1962: 217–225). According to Barth, geographic and social isolation could not explain the persistence of cultural difference among cultural groups who had knowledge of and social interaction with one another. Nor did social interaction dissolve ethnic distinctions. Barth's work showed the opposite process; instead of diminishing or diluting ethnic distinctions, social interaction seemed to activate a conceptual distinction between members within and outside the group. In contrast to the assumed processes of acculturation, ethnic boundaries were actually created through contact. This comparative process illuminates cultural differences through a dialectic of comparison between members of dichotomized groups. Following the idea developed by Park, Gluckman, and Mitchell, the signals of inclusion and exclusion were defined as "diacritica." These signals were both material and behavioral:

The contents of ethnic dichotomies can be of two orders:

> i. overt signals or signs-the diacritical features that people look for and exhibit to show identity (dress, language, house form or general style of life) and,
>
> ii. basic value orientations: the standards of morality and excellence by which performance is judged. . . . Belonging to an ethnic category implies being a certain kind of person, it also implies a claim to be judged and to judge oneself by those standards that are relevant to that identity. (Barth 1969: 16)

These diacritica reflect a shared consensual and conceptual vision of what constitutes "proper" behavior for a person of that ethnicity. Choices in dress, language, house form, marriage partners, and technology were mediated and referenced through the social unit a person belonged to. The acculturative processes assumed by many archaeologists and that generally characterize contemporary approaches to contact period studies imply a weakening of boundaries through increased interaction. In focusing on the boundaries between groups, Barth was confronted social processes that radically challenged the definitive view of culture change. These notions are central to the development of Indigenous archaeology in that they provide a more complicated means of assessing the relationships between a material past and contemporary Indigenous communities. Acculturative models assume that continued contact between groups "require and generate a congruence of codes and values." Barth's work demonstrated that the recognition of difference and the generation of ethnic boundaries appeared to have the opposite effect. Situations of contact and interaction (what group is not interacting?) actually elicit socially exclusive behaviors that restrict interaction and buffer cultural systems from outside influences. These boundaries determine the points of contact between different cultures and functionally preserve cultural differences deemed relevant to the group. The notion of the perpetually vanishing primitive is an essentials rendering of Indigenous culture and is quickly exposed as such when these concepts are examined more closely.

Closely paralleling the systems and domains of interaction outlined by the Chicago school's social distance scale, interactions between groups were mediated through culturally defined standards of behavior. Being identified with a particular group implied a "shared criteria for evaluation and judgment . . . a potential for the diversification and expansion of their social relationship to cover virtually all sectors and domains of activity" (15). The opposite process placed limits on the kinds of interactions which one could engage in: "the dichotomization of others as strangers, as members of another ethnic group implies a recognition of limitations on shared understandings, differences in the criteria for judgment of value and performance, and a restriction of interaction to sectors of assumed common understanding and mutual interest" (Barth 1969: 15).

As a behavioral principle, ethnicity affirms membership in a group and defines others as outsiders. It is only one facet of an individual's identity that can include age, gender, filial, clan, and kin associations but, most importantly, provides the context in which these identities are expressed, recognized, and given meaning. From the perspective of the individual, being a member of an ethnic group reflects a consciousness of difference and "implies the cultural structuring of the social universe" (Comaroff 1987: 303). This "structure" provides the context in which human agency is expressed. As Barth states:

> Since belonging to an ethnic category implies being a certain kind of person, having that basic identity, it also implies a claim to be judged, and to judge oneself, by those standards that are relevant to that identity. . . . If they say they are [in group] "A" in contrast to another cognate category "B," they are willing to be treated and let their own behavior be interpreted and judged by A's and not B's; in other words, they declare their allegiance to a shared culture of A's. (1969: 15)

As such, ethnicity is superordinate and serves as the basis for any number of social personalities an individual may assume, each one having the potential for individual or collective social action. This is a key element of ethnicity theory and helps move the concept from a purely psychological phenomenon to a social process that regulates and shapes interaction and behavior. This approach locates ethnic groups on an equal footing with other forms of social organization (e.g., kin, clan, family, and village). Defined in this manner, ethnic groups have organizational capacitates that are capable of generating large-scale social movements such as the Revolt of 1680.

Despite Barth's important theoretical breakthrough, there remained definitional inconsistencies that made discussions of ethnicity particularly difficult. Ethnic groups, unlike the tidy and bounded cultural units that ethnographers described, were ephemeral, shifting, and inconsistent. They changed and were transformed through time, and most troubling, they could be either more or less exclusive depending on the context. This ambiguity created several important questions: If ethnic identities were deeply rooted within all cultures, did they share a common set of attributes? If ethnicity were a largely a ephemeral and cognitive process by which individuals created order out of a socially complex world, how could it possibly be studied? How were these purely psychological social categories, which passively regulated interaction between groups, transformed into powerful and sometimes violent social movements?

Viewed as a central concept in Indigenous archaeology, ethnicity has a multiplicity of applications both ancient and modern, psychological and global. Much of the confusion in debates surrounding ethnicity result from the attempts of archaeologists to apply a single model of ethnicity to all situations at all times.[9] As historical contexts change, so do the character and applicability of the concept. From the perspectives of Indigenous archaeology, ethnicity is best understood as a social process with different qualities at different stages of development (Voss 2008). Following this more processual approach, more productive analyses of ethnicity concentrate on the circumstances that generate ethnic identities, produce conflict, and effect change between ethnic groups. These conflict theories serve as the model for my interpretation of the events surrounding the Pueblo Revolt. This evaluation of ethnicity, boundary maintenance, and ethnic conflict is based on the following assumptions:

- Ethnicity, in its broadest sense, refers to the sociology of group interaction. It is a form of identity, both collective and individual, that is grounded in a concept of group affiliation; as such, ethnicity is a form of social organization.
- All societies are plural. All peoples exist in a universe populated by different groups, and each acknowledges, defines, and ultimately creates group membership in relation to other groups.
- This ascriptive and conscriptive mode of classification is a universal process. This associative process relates the individual to the group and the group to the individual.
- While structuring interaction between groups, ethnic identities are, at some level, subject to human agency; individuals can attempt to transcend the social structures that relate the individual to the group as well as the interactions between groups. In this sense, ethnogenesis can be defined as a form of "social practice" (after Giddens 1984).

As a form of social organization, ethnic groups have the following characteristics. The scale of these characteristics as well as their specific contents are determined by the particular groups in contact with one another. Contexts, historical and cultural, are crucial elements in any study of ethnicity. Membership in an ethnic group invokes a concept of social distance and suggests a common regional or geographic origin. These common origins invoke either fictive or biological kin relations as the basis for group membership. This implies an ancestral connection to others through biological or fictive kin. In this sense, like Benedict Anderson's imagined communities (1985), ethnic identities have a transcendent quality and "extend" the life of the individual into the future and the past through descendants and ancestors.

As such, ethnic groups reference a collective (or exclusive) past through collective memories, codified in origin myths and oral histories. Because it is provided by the group, ethnicity is a superordinal feature of an individual's identity, and it implies an allegiance to a set of values and standards of performance and judgment. Linked in this way to status, it provides the context in which other aspects of identity (including gender, age, and filial and clan relations) are recognized and evaluated. Because of these characteristics, ethnic identities have certain organizational capabilities. As I argue later, affiliation with the group conditions social action and provides the basis for collective behavior and intergroup relations. These collective actions include mass migrations and abandonments, resettlements, and group violence. Each of these processes can be observed and documented in the years leading to the Revolt of 1680.

This processual approach to the study of ethnicity does acknowledge the limitations associated with archaeological practices. First, categories of diacritica comprise a list of possible behavioral markers that separate communities. Without extensive contextual evidence including historical texts, ethnographic information, and archaeological data, the conflation of diacritica with a set of empirical features is speculative. Based on a diachronic or processual model, as groups change, so do the specific signals that identify those groups.

Diacritica: The Signals of Social Distance

As symbols that signify membership in a community, diacritica can be somatic, behavioral, or performative. Isaacs's discussion of "the group" (1975: 31) as the setting in which an individual acquires an identity is particularly instructive. The group provides an individual with "ready-made endowments" that relate the individual to the community in various ways;

and biological inheritance, group socialization, rites, and rituals provide the structural framework for the development of ethnic identities.

Despite Barth's emphasis on the situational and fluid character of ethnic boundaries, the ability of the individual to assume any ethnic identity is limited by the characteristics of specific diacritica and by the ascriptive criteria of other groups. This is particularly true in cases where there are imbalances in power between ethnic groups. Specific somatic diacritica, such as skin color, are nontranscendent. Correspondingly, the most stable and least flexible are somatic features. These include skin color, hair texture or color, physique, and eight. Body modifications such as scarification, hairstyles, teeth filing, and circumcision also serve as meaningful markers of group membership. Proscriptions that prohibit intermarriage between groups tend to reinforce the salience of these somatic features. Because of their rigidity, these diacritica often serve as markers in highly stratified multiethnic societies; any attempt to assimilate through behavioral or performative means will be impossible. As I demonstrate in later chapters, the creation of racialized social castes was a crucial element in Spanish colonial society.

Behaviorally, linguistic socialization provides symbolic and nominal referents. Individual, family and group names relate the individual to members of the group as well as to individuals outside the group. More subtle inculcative features such as syntax, accent, gestures, and speech mannerisms emphasize distinctions. Preferences and tastes are reinforced through choices in clothing, body modifications, and grooming. Group performances reinforce distinction and affiliation through rites of passage and collective and individual ritual; other forms of discourse recall and reproduce collective memories through oral histories and origin myths. These also relate the individual to the geographic and spiritual landscape and define the social distances between groups. As we shall see, mythological kinship structures serve as an important vehicle for ethnic movements. The following discussion of ethnic subordination, stigmatization, and ethnic conflict will demonstrate the importance of these criteria in maintaining structural imbalances of power between groups.

Conflict Theories: The Source of Resistance and Change

A common theme in processual discussions of ethnicity is the idea that power imbalances between parallel groups transform these latent, organizational categories into powerful social movements. The study of *ethnogenesis*, defined as the process by which ethnic groups are created or transformed, has received little attention in archaeological literature. Synthetic analysis by Cohen (1977), Comaroff (1987), and Horowitz (1985) have expanded the discussion of ethnicity to include a detailed and comprehensive analysis of the circumstances in which ethnic groups emerge.

According to Comaroff (1987: 307), ethnicity has its origins in the asymmetrical incorporation of structurally dissimilar groupings into a single political economy; as long as no group is ascendant, symmetrical or parallel relationships are perpetuated. Parallel groups do not have to be peaceful neighbors; they can exhibit open hostility or mutually beneficial relations, but when a group is able to exert its dominance over another and removes from it the means by which value and labor are given meaning, ethnic conflicts result.

Parallel group interaction can be transformed to ethnic conflicts by a number of different processes. Expansion, migration, and conquest all have the potential to change the relationships between groups or bring formerly segregated groups into contact with one another. The colonial movements that followed the exploration of the Americas involved each of these elements.

CREATING THE INDIAN: COLONIZATION AND ASCRIPTION

Ethnic identities are grounded in historical patterns of interaction with other groups. Colonial expansions during the contact period extended the boundaries of Spanish society to include new geographies and communities within a single sphere of interaction. These groups were defined by the forces that motivated this interaction; the objectives of religious conversion and economic extraction were embedded the Spanish classificatory system. New ascriptive categories such as "Indios" or "Negros" were created by bureaucratic fiat. Eric Wolf in *Europe and the People without History* (1982) suggested that these ascriptive criteria are created through economic expansion and mercantilism.

According to Wolf, the function of these categories is exclusionary and creates differential rights and obligations within the system. These general categories incorporate local identities within a larger unit and force the adoption of a new more general identity, with lower status. Groups were rated in terms of their social, geographic, and biological distance from Spain. The linguistic correlates of this process reinforce a symbolic and moral distance between groups. The names of Spanish ethnic categories, *Negros* and *Indios*, disregarded any cultural or physical distinctions among colonized peoples and denied "any constituent group a political, economic or ideological identity of its own."

STIGMATIZATION AND SUBORDINATION

Ethnocentrism is a crucial element in the maintenance of ethnic boundaries. The imperatives of colonial empires were based on an assumed superiority in all facets of life. Without this crucial element, the basis for colonial authority would dissolve (Horowitz 1985: 160). The establishment of social distance and the conflation of morality with oneself and immorality with "the other," then, serves as a justification for the treatment of other people as nonhuman or less than human. Stigmatized groups can experience segregation, discrimination, and persecution based on their ascribed identity. Despite the fact that these subordinate identities have little basis in any preexisting social reality, a common status is reinforced by the dominant group through universal restrictions (Horowitz 1985: 26). Spanish colonial societies imposed restrictions in dress, language, marriage, access to education, and occupation among Indian communities. Ultimately, the objectives of these categorizations involved the consolidation and maintenance of power and the control of resources by the nonstigmatized, dominant colonial strata of society.

According to Cohen (1978: 391), resources are defined as "the instrumentalities used to satisfy culturally defined needs and desires." For Indigenous subjects, these included food, water, land, and labor, as well as spiritual, technical, or cultural information; colonial power rests in the ability to control and allocate these resources. One of the ways in which colonial societies were able to exert their authority over Indigenous subjects was through leadership replacement and selection. Spain, like many colonial empires, employed a combination of direct and indirect rule in the maintenance of power. The identification and recognition of cooperative Indigenous elites restructured the manner in which resources were allocated and determined the manner in which power and authority were expressed, acknowledged, and legitimized through ceremony, tribute, and ritual. This act of intervention threatens a whole other realm of Indigenous social integration in which status, authority, and identity were recognized and given meaning.

This dyadic social structure, institutionalized through force, produced a more fluid social dynamic. In these situations, individuals can choose to invest themselves (via identity) in an alternate and sometimes antagonistic status positions (Indios, Negro,s or Ladinos). This ambiguity and fluidity create tensions between and within the systems as the symbols and practices that clarify Indigenous membership are altered. Three responses follow: assimilation, migration, and confrontation. Each characterizes the interactions between Spanish colonizers and Native peoples in colonial New Mexico.

Responses to Subordination: Assimilation and Conflict

> Neither ranked or unranked ethnic systems are static. Among the engines of change is the ethnic conflict itself.
>
> Donald Horowitz, *Ethnic Groups in Conflict* (1985: 32)

> Once upward mobility is seen as a possibility and energies are expended to achieve it, ethnic groups must inevitably become internally stratified.
>
> John Comaroff, "Of Totemism and Ethnicity" (1987: 315)

To the extent possible, some colonial subjects respond to changes within this social universe by attempting to emulate the behavior or assume attributes of the dominant group. This emulation can take several forms. Individuals may pose as defectors and adopt new behaviors to better monitor the activities of the dominant group; or, as marginal or alienated members of their original communities, they may seek to elevate their own status within the new system in opposition to those who do not make this attempt (Horowitz 1985: xx). Frequently however, the cost of defection is steep. These status positions, based in opposition to the interests of one's kin groups, are often tenuous. The ascriptive diacritica of the Spanish were the highly stable, largely somatic attributes of "race" (Elliott 1998). In the rigidly stratified society of the Spanish Empire, individuals were largely unable to transcend the structural limitations of these categories no matter how they changed their behavior.

Although these ascriptive qualities may not signify membership in any group the subjects recognize as valid, as these diacritics are used as the basis of differential treatment, the opportunities for conflict between the colonizer and the whole suite of social groups will increase. Some groups will attempt to remove themselves from the social system in which these inequalities arise. Mass migrations and refugee movements (both of which characterize the contact period in New Mexico) are one possible course of action. If (physical) group mobility is restricted, the differential behavior toward stigmatized groups tends to generate a solidarity based on the common experience of discrimination. Structural inequalities based on the diacritics of the dominant group generate responses of a similar scale. As an unintended consequence of subordination, the collective experience of persecution may reinforce the relevance of the ascriptive category and motivate collective actions that attempt the remove the source of discrimination (Comaroff 1989: 312).

These ethnic conflicts or "cultural movements" (Horowitz 1977: 6–18) adopt, articulate, and codify the shared diacritics of the subordinate group. These diacritics become emblems and symbols that serve to further segregate the integration and interaction of the two social domains as individual behaviors are more carefully scrutinized. "Group identity is thus infused with a new or revived cultural content that serves to demarcate the lines between the groups more clearly, reducing the ease with which members can cross boundaries" (Horowitz 1985: 72). Languages and religious practices are "cleansed," and group members assert a return to orthodox practices that existed prior to contact with the dominant group. In colonial New Mexico, the Revolt was of 1680 preceded by a similar set of invocations by the Pueblo leader Po'pay.

Here kinship terms were deployed (Horowitz 1985: 57) and served as the organizational framework for the pan-Puebloan movement. This more general pan-ethnic identity incorporated local identities without eliminating their cohesion or identity. In the processual model of ethnogenesis, origin myths and oral histories serve as vehicles for these changes in a relational discourse. Regional kin networks help facilitate and evoke rights and obligations that crosscut and subsume traditional rivalries and animosities. At this point, there may be sufficient momentum to facilitate an organized movement of rebellion. Regardless of its success or failure, social movements based on these identities reinforce the salience and symbolic unity of the pan-ethnic identity (Comaroff 1989: 316). These actions are then incorporated into the regional and local histories of constituent communities.

The creation of these movements demands new symbols that invoke the complementarity of the various groups. Disputes over choices in leadership and the specific version of orthodox practices may reactivate tensions that existed prior to the arrival of colonists. Because no group is completely ascendant, the more inclusive, emergent pan-Indian identity may subsequently be threatened by significant cultural deviations as some of the merged sub-groups discover, at closer range, traits among other subgroups that they regard as "highly unusual or in need of change" (Horowitz in Glazer and Moynihan 1975: 124). The activation of pan-ethnic identity may generate new diacritica and symbols that may not be able to sustain the movement beyond the resolution of the crisis.

At the same time, this activation may elevate the status of individuals in ways that are not beneficial to the other members of the group. Pan-ethnic elites may assume positions of authority that run counter to the interests of the individual subgroups; a lack of confidence in the leadership of these movements can undermine the strength of the movement and create factions. Leaders must be trusted to act in the best interests of the group rather than some subsection of it with which they are identified (Cohen 1978: 397).[10] If the members are successful in removing the source of inequality, these movements frequently splinter, and local identities resume as the ascendant forms of social organization.

For the dominant group, the threat of future collective action changes the terms by which the groups interact and allows for the incorporation of social practices introduced by colonial societies. In this sense, the outcome of colonialism is not a monolithic replacement of cultural systems through military conquests and "conversions." This dynamic helps explain the presence of colonial institutions, religious practices, and materials among contemporary Pueblo communities and provides a more contextual, multivalenced interpretation of that presence. Local versions of Catholicism, the compartmentalization and isolation of ritual practices, and the construction of identities that actively articulate cultural distinctions are reiterated in performances such as the Feast Day and katchina dances. Within this context, these performances acquire a new significance and meaning.

With this more nuanced understanding of the processes of cultural interaction provided by an application of ethnicity theory, we are able to enact and conceptualize an approach to colonial contacts that simultaneously explains the presence of Puebloan traditions in contemporary society and deconstructs the traditional historical narrative of Indigenous cultural collapse. In deconstructing these narratives, an Indigenous approach to archaeology seeks an interpretive framework that provides the means for a more inclusive rendering of colonial history and helps create a space for Indigenous voices and perspectives. From this vantage, we are led to examine primary historical documents of the sixteenth century with a different set of assumptions. Rather than reiterating the often simplistic account of Indigenous cultural collapse via acculturation, missionization, and military conquest, Indigenous archaeology facilitates an excavation of colonial mythology. This discussion provides the historical and cultural context in which interaction and contact between the Pueblos and Spanish colonists and missionaries took place. The processes of pan-Indian ascription and subordination, and the emergence of pan-Puebloan ethnogenesis, are more clearly observed through an analysis of primary historical texts, ethnographic accounts, and archaeological materials. In the second half of this book, I correlate these contextually situated processes with occupational shifts, migrations, and the development of new kinds of communities within the Pueblo world.

CHAPTER 4

The Mythologies of Conquest

Militarizing Jesus, Slavery, and Rebellion in the Spanish Borderlands

In the immensely popular book *Guns, Germs and Steel: The Fates of Human Societies*, ornithologist Jared Diamond introduced what has become (like it or not) the most widely read conquest narrative ever written. Published in 1995, *Guns* argues that contemporary disparities in power, capital, and wealth between the "developed" and "undeveloped" world, between Euro-colonial powers and Indigenous peoples, emanate from a series of geographic, technological, and historical accidents. Disguised as an attack on racial determinism, Diamond's book argues that the conquest of the Americas and the dispossession of Indian wealth resulted not from the innate inferiority of Indigenous peoples but from the fortuitous geographic position of Europe at an axis in which superior military technologies (guns and steel) and immunity to diseases (germs) accumulated.

Diamond's work relies on much of the same logic and interpretation present in the mythology of the invisible Indian. His books demonstrate with remarkable clarity the role that disciplinary boundaries have played in the construction of terminal narratives. Diamond has helped generate a more sanitized and palatable version of conquest—a new revisionist strain in which the colonization of the New World hinged on a series of fateful battles, where diseases led to conquests and where conquest itself is reformulated as a kind of accident of military technology and natural selection. This new interpretation speaks to a general discomfort (and outright rejection) many contemporary readers share over the use of state- and religiously sanctioned violence against Native peoples.

Appealing to the sensibilities of a more critically reflective audience, Diamond rejects the notion that Europeans were intellectually superior to their Indian foes (racial determinism) and suggests that that the use of gunpowder and steel—technologies that fortuitously accumulated in Europe—allowed small groups of highly outnumbered Spaniards to conquer a numerically superior foe. In this reworked version of conquest, biology is the real culprit. European diseases helped clear the landscape of an Indian presence and enabled the installation of new regimes that quickly appropriated "unclaimed" and "undeveloped" Indigenous wealth. In Diamond's new mythology, geography, biology, and technology were the real determinants of colonial conquests and European dominance, not racial superiority.

The narrative has a powerful allure. Diamond argues that *any* human group so situated would have conquered its neighbors. What Diamond's thesis does not explain is the central importance of human agency in any of these events. Conquests, wherever they happen, are never accidental. They require ideologies that help initiate, sustain, and justify the use of violence against other peoples. Without a belief system that situates an individual as a member of a group and that locates that group in relation to other groups, conquests are impossible. Nowhere is this idea more powerfully demonstrated than in the story of Spanish colonization.

The creation of the Indian as a single subordinate class of human, the justifications for the use of violence against those peoples, and the dispossession of Indian wealth were firmly grounded in the philosophical traditions of early modern Christianity (ca. 1500). In the system of Spanish colonial expansion, the development of an aggressively evangelical, militant strain of Christianity was of far more importance than some historians have been willing to admit. Population levels were greatly affected by slave raiding, the forcible use of Indian labor in cash (rather than food) crop cultivation, and the enforcement of unsustainable tribute demands by Spanish colonists. Each of these activities profoundly compromised the health of Indian communities. The effects of stress and starvation on human health are today undeniable. Agricultural economies require surpluses and stable sedentary populations to tend fields, manage water resources, and exploit supplementary sources of food. Population declines (where they did occur) were caused by human agents, not free-floating viruses in a cosmic lottery.

The same is true for the use of military technologies. In most cases where Spanish soldiers are credited with extraordinary valor in their encounters with groups such as the Aztecs (accounts that were frequently written by the soldiers themselves), thousands and thousands of unmentioned Indian allies simply overwhelmed the target populations. These facts have only recently made their way into secondary historical literature. Technological advantages were often precarious; early firearms were both cumbersome and difficult to load, and powder was vulnerable to the 1,300-mile journey from Mexico City to Santa Fe—roughly the same distance from Madrid to Kraków, Poland. Should a mishap occur, supply lines were vulnerable and poorly mapped—reinforcements were months away.

Where the Spanish did enjoy an advantage was in the total and unremitting brutality of their tactics. Often after an initial skirmish ended, soldiers would corral their surviving foes and then burn them alive, execute them, or maim them. In the Tiguex War (1540–1541) waged by Coronado, colonial soldiers used these tactics at several villages—a pattern that was repeated during every subsequent expedition. There is no evidence that the Pueblos engaged in this type of warfare prehistorically, nor would they have expected the total war of fire and blood (*un guerra de sangre y fuego*) that was visited upon them. The concepts of *jus ad bellum* (Latin for "justice to war" or "right to wage war") and total war were carefully crafted by Christian theologians. This concept of intimidation through violence is mentioned throughout the colonial period as a tactic of social control. Technology may have made this easier but is in itself not an explanation.

Ideologies that allow one group to imagine itself as the naturally or divinely ordained instrument of salvation, progress, and change are human constructs enacted by individuals. The philosophical justifications for the subordination of any people are not only the most fundamental elements of conquests but also the most enduring and interesting. What greater irony could one imagine than the manner in which Christianity, which begins as a small sectarian strain of Judaism rejecting materialism and the accumulation of wealth and embracing a radical form of spiritual egalitarianism in the space of a few centuries, is yoked violently to the arms and economic ambition of the Spanish state?

Conquests are frequently authorized by religious edict or state proclamations. These are events in which the normative codes of morality are formally suspended and where the individual imagines him- or herself (and assumes the position) of a naturally or divinely ordained instrument of salvation, progress, and change. The use of violence in this case does not result from any flaw in Spanish or Hispanic character (or the ridiculous stereotype of hot-blooded peoples) but from the application of a carefully reasoned, state-sanctioned code of behavior. All of the states in early modern Europe deployed these legal fictions in the subjugation of Indigenous peoples—Spain just kept better records.

Economic pressures in the home country and the desire for new sources of bullion, Indian labor, and tribute facilitated the development of a lawless, poorly administered colonial culture in the northern frontier of New Spain. Any encounter with the primary historical documents—letters between officials, descriptions of encounters with Indian peoples, official inquiries, and court testimonies—reveals the degree to which violence was routinely deployed (even celebrated) as a tool of intimidation and control.

As the first truly modern national empire, Spain struggled with the limits imposed by its economic, political, and bureaucratic resources. After the initial flush of wealth provided by the conquest of the Aztec and Inca empires had been exhausted, the search for silver mines in the northern frontier intensified as bullion became an increasingly vital source of wealth for the home country. The search for profitable and self-sustaining mining ventures led to the exploration of New Mexico. Of paramount importance in these encounters was the exploitation of Indian labor. Colonial ventures had to be self-sustaining, or the licenses for colonial settlements would not be granted by the Crown. These offices were often financed with monies borrowed from other, more wealthy colonial benefactors. Unless a conquistador could prove that he and his followers (often numbering in the hundreds) could be supported by Indian labor, and unless these fees were paid back through the collection of additional tribute, colonial expansion was doomed to financial failure. This explains the impetus behind the wildly inflated population estimates of early expeditions (and often cited by disease supporters). If these pressures were not significant enough, a parallel society of missionaries would compete with secular officials for the same limited resources of tribute, food, and labor.

The financial situation of the Spanish state was similarly precarious. While the fall of the Aztecs in 1519 seemed to initiate an instantaneous transfer of power and a limitless stream of cash to the home country, military encounters on the northern and southern edges of the Spanish Empire were both extremely costly in human and material terms and far more ambiguous by any measure of success. Lacking the large sedentary populations of the Valley of Mexico and faced with the highly inflated costs of finished goods (which the colonists were required to purchase), colonists on the frontier came to rely on an illegal yet pervasive Indian slave economy to provide a cheap labor supply. Coupled with the forcible conversion and eradication of Native religions in those regions, the Native peoples all along the northern frontier staged a series of successful rebellions that were to last well into the seventeenth century. In fact, the Pueblo Revolts of the 1600s were not unique. The cycle of slave raiding, resistance, rebellion, and violent suppression by colonists in Nueva Vizcaya established the tactics, motives, and justifications for the subordination of Puebloan peoples.

Like their southern neighbors, the Pueblos quickly realized that the only hope for their survival was to vacate, temporarily or permanently, their settlements—to impose geographic distance in order to maintain social distance, to limit interactions with and exposure to colonial violence through mobility. Abandonment was a social strategy used to ensure survival. And disease was not the culprit. This response, largely ignored in the interpretations of many historians and archaeologists, calls into question the idea that military encounters (Spanish or otherwise) facilitate the instantaneous transformation of Indian cultures, religions, and economies. In fact, as the history of the northern borderlands demonstrates, these actions often produced the exact opposite result. The common experience of violence, the suppression of religions, and the disruption of subsistence practices invoked powerful, countervailing pan-ethnic, fundamentalist movements. The histories of these movements should raise questions about disease-based explanations of abandonment and the meaning of conquests in general.

The Power of Naming: The Birth of the Spaniard and the Indian

Colonial expansions frequently initiate processes that can lead to ethnic conflicts and pan-Indian and fundamentalist movements. Just as the process of ethnic ascription and subordination help determine the relations between ethnic groups, so do the economic and social motives that generate these contacts. Emerging in modern Europe in the sixteenth century, nationalist identities manifest and exhibit many of the same organizational properties and characteristics as ethnic groups. The primary difference between nationalist and intraethnic encounters is that with modern colonization, Europeans were able to draw from the almost limitless material, technological, and financial resources of the state. Through written records, they maintain the power to recognize or ignore, to acknowledge or dismiss, the diversity that exists on the ground. Because nations benefit from colonization through the expansion of resource bases and taxation, nationalism, capitalism, and colonialism tend to reinforce one another. Spain's economic and social policies demonstrate this process quite clearly. The same cannot be said for ethnic groups or Indigenous movements in general. Because there is no formal institutional apparatus for the accumulation of capital, the transformation of corporate wealth into the machinery of warfare is frequently impossible.[1]

Colonized groups were rated in terms of their alleged social distance from Spain; linguistic discursive practices reinforced the putative symbolic and moral distance between Europeans and Indians. The naming of stigmatized populations as "Negros" and "Indios" simultaneously disregarded the cultural or physical distinctions among colonized peoples and imposed social categories based on the diacritica relevant to Europeans. The process of pan-ethnic ascription, based on the European diacritica of "race," relegated stigmatized populations of Negros and Indios to lower levels in the labor market and insulated colonizers from competition from below (Wolf 1980: 380). Because the expansion of colonial empires presupposes and depends on unequal access to resources, racially stigmatized populations were unable to transcend the limitations imposed by their own physicality. No Indian or African could ever *become* a Spaniard. While acculturation models assume universal access to symbolic and material resources, in colonial societies, full assimilation is a fiction.

But the rigidity of the Spanish colonial system was neither uncontested nor intractable. While some members of colonized tribes attempted to emulate the behaviors of Spanish colonists and improve their own statuses, the cost of defection was often steep. The new status positions, based in opposition to the interests and social identities of their own kin groups, were often tenuous or even dangerous. These individuals are highly valuable to colonists and highly vulnerable to retribution when the colonial system breaks down.

The systems of ethnic identification in colonial Spain were sustained by structural inequalities that defined Jews, Muslims, and Christians. Although religious conversion allowed for the partial integration of Indians into the colonial social system, ultimately full access to the whole range of social identities and statuses within Spanish society was thwarted by the diacritica of race.

The Spanish crown was able to project this system throughout the Americas. While the viceroys of Aragon, Italy, Portugal, and Flanders were ruled by the elites of their respective groups, the viceroys in the Indies were drawn exclusively from the ranks of Castilian nobility (Elliot 1969: 172–177; Anderson 1983: 56). Early attempts to establish a Native priesthood (considered too risky by colonial authorities) were abandoned by 1555. Access to advanced education and religious texts were similarly limited to Spanish colonists. The Colegio of Santa Cruz Tlateloco was established in 1536 to train the sons of Indian nobility to serve as the next generation of Indian teachers (Bernardino de Sahagún, the individual most responsible for the preservation of Aztec historical texts, taught here), but the school was closed in 1555 when laws were passed forbidding the creation of a Native clergy (including nuns), as well as the equal participation of Indians in government. Initially, Indian education was limited to males of the upper classes, but the universities at Mexico City and Mérida taught Nahuatl and Otomí to Franciscan missionaries (Meyer 1999: 149). The Spanish system ensured the maintenance of social boundaries through these institutions. The language of state was Castilian, but the sacred language—the language of doctrine, law, and the gospels—was further insulated from Indians through the use of Latin. Less obvious than guns or steel, these institutional barriers sustained inequality and maintained conquest in bureaucratic silence.

Status was often determined through a complex system of racial gradients and classifications (Meyer 1999: 203–219; Miller 1985: 138).[2] This system allowed for the partial integration of select populations within the colony and ensured the exclusion of others. In the early stages of colonial expansion in New Spain, Tlaxcallan, Tarascan, and Aztec allies were often deployed as soldiers or agents of colonial authorities. In exchange, they were exempted from many of the formal restraints and tribute obligations of other groups (Miller 1953: 34; Meyer 1999: 199). Although *caciques* and *fiscales* (Indian secular and religious leaders) were appointed as representatives of colonial authorities in Indian communities, their authority was compartmentalized and limited at a local level. Unlike English and American colonial systems, forms of partial citizenship were freely extended to Indians. In regions close to colonial capitals, separate court systems were established to hear the complaints of Native peoples (Borah 1983: 79–120). But as the following discussion demonstrates, in the frontiers of colonial New Spain, many of the laws and institutions designed to limit the abuses of Indian subjects were impossible to enforce. The much-heralded New Laws (forbidding the enslavement of Indians) were passed in 1542, but thirty of its provisions were repealed by Charles V in order to stave off the rebellion of his subjects in Peru and Mexico.

Colonial systems are highly vulnerable to the social changes they unleash. The creation and imposition of subordinate social categories often lead to unintended consequences for the dominant group. Tactical victories by military forces, the use of violence as an implement of social control, and the cessation of overt resistance do not signal a wholesale or immediate subjugation of "conquered" peoples. Military conquests, often depicted by colonial historians as demonstrations of European dominance, frequently imply a transformative significance within colonial narratives, but they do not account for more pervasive and effective tactics of Indigenous resistance. The often heavily valorized accounts of victories by Pizarro, Cortés, and Coronado serve as evocative symbolic events for Europeans, but they generate a completely different meaning and signification among Indigenous peoples. In fact, these acts of violence, coupled with the common postevent experiences of discrimination, segregation, and persecution, often help generate collective responses by subordinate groups. Etched in the memories of subject peoples, these actions serve as potent and enduring vehicles for the development and implementation of more subtle and powerful forms of opposition. The Revolt of 1680 may represent the culmination of one such resistance movement, but it is part of a much larger process of negotiation, defiance, and organized conflict by Indigenous peoples.

Creating Indian and Spaniard as Social Categories

In colonial Spain, contact between the Pueblos and the Spaniards coincided with the development of Spanish national-ethnic identity and the advent of a capitalist world economy. As contact with the known world expanded to include previously unknown geographic regions and peoples, the social categories that defined and divided Europe were projected on the Americas. The Indian came to symbolize a new kind of person singularly defined by geographic isolation from Europe and Christianity (Williams 1990: 13–48). For many Europeans, the presence of peoples outside the historical religious and philosophical traditions of Europe had deep symbolic significance. By virtue of his isolation, the

Indian was conceived of as a being who demanded transformation in order to be included within the realm of other humans. If the Indian had been prevented from hearing the gospels of Jesus Christ through geographic distance, Spanish colonization provided the instrument of salvation. That same logic was to doom an entire continent of African peoples to the yoke of legalized slavery. If Jesus and his message could have reached African tribes through geographic links, the continued practice of non-Christian religious traditions (paganism) could theoretically be understood as a rejection of Christianity. Rejection of Christ, however imagined, was grounds for enslavement according to canon (Christian) and Spanish law (Williams 1990: 1–118). These legal and philosophical systems helped buffer colonists from the brutality of their actions by invoking a rationalizing mythology of violence in the service of God. Jesus had been militarized.

In contrast to both "black" and "white" revisionist histories (which make no mention of the legal basis for the use of force against non-Christians), conquests, slavery, and social hierarchies based on religious and racial categories were an integral part of late medieval Spanish social and political life. Serving as the primary justification for conquest, colonization, and forcible conversion, Spain's missionary efforts generated both debate and dissent between the Spanish kings and the colonists of the New World. But the rules of conquest and the spoils of war had been clearly delineated during the Crusades in the Near East as well as in Spain. As established by the *Reconquista* (Reconquest) and the expulsion of the Jews during the fifteenth century, the rejection of Christian faith by Native peoples justified the use of "complete," "total," or "just" wars. Despite the rhetoric of Black Legend and White Legend revisionist histories, the deployment of violence as a means of establishing hierarchical relations or dominance over Indian subjects was neither rash nor impulsive. The use of force as a tool of subordination was carefully researched and debated by secular and religious authorities throughout Spain.[3] The invocation of the Black Legend confuses a critique of Spanish colonial policies with attacks on Hispanic identity—completely ignoring the fact that Indigenous peoples, Africans, and mestizos are also Hispanic.

Spain suffers mostly from the fact that the Crown required meticulous record keeping; it encouraged accountability and power struggles between and among its colonial elites (all of which were dutifully recorded), between secular and religious officials. Spain actively prosecuted colonists guilty of war crimes against the Indians. None of these institutional apparatuses was a part of English colonization. Accomplished through the development of quasi-governmental institutions buttressed by English military power, the first global corporations—the Massachusetts Bay Colony, the Virginia Companies of Plymouth and London—were privately financed. No public records were required to be prepared or disclosed. These fictional institutions, imagined in boardrooms and invented on paper, were not required to consider any moral obligations or standards of behavior with regard to Indian peoples. Corporations have no Bills of Rights. While initial contacts between Spanish colonists and Indians were often bloody and lasted for decades, the Crown was comparatively very aggressive in its defense of Indian rights and the persecution of offenders. English and American philosophies of colonization were equally brutal but afforded none of the same protections.

At its peak, the Spanish Empire was massive and diverse. Yet the administration of the Indies was of particular importance and benefit for Castilian nobility. With the ascension of Charles V to the Spanish throne in 1516, Spanish possessions included not only the Americas but also Naples, Sicily, Sardinia, the Netherlands, as well as kingdoms in Germany and France. As heir to the Hapsburg dynasty, Charles V's claim as Holy Roman Emperor and "defender of the faith" extended the influence of Spain within Europe. At the same time, the Catholic Church, threatened with defections from a growing movement toward Protestantism, sought to reassert its authority with the aid of Spanish military might. In sponsoring religious crusades against the Moors in Spain's southern provinces, Ferdinand and Isabella had extended the power and prestige of the Crown and opened territory and resources for Castilian nobility. Marking a shift in power from a feudal, provincial system of self-governance, the authority of the Crown was sustained through a rapidly evolving, centralized, and extensive royal bureaucracy. In newly conquered territories, the crown was able to increase taxation, institutionalize Castilian as the official language of state, establish uniform legal codes, and consolidate its own authority vis-à-vis competing provincial nobility.[4]

The conquest of the Aztecs and Incas helped extend the power of Spanish monarchs throughout Europe. In this sense, the expansion of the Spanish Empire was an integral component in sustaining the political objectives of the home country. While the rise of Protestantism in Europe challenged the hegemony of Catholicism (Lutheranism was established in 1521, Anglicanism in 1534, and Calvinism in 1536), the exploitation of new resource bases and the taxation of economic activities supported and reinforced the political objectives of the Spanish monarchs in Europe.[5] These events coincided with what is commonly recognized as the most corrupt period in the history of the Catholic Church. The sale of indulgences, the initiation of massive and costly artistic ventures (St. Peter's basilica in Rome and the Sistine Chapel were commissioned at this time), and the moral decline of the clergy would lead to an internal Catholic revolt: the Reformation (1517–1648). Lacking the arms of the emerging Spanish state, the Catholic Church installed Charles V as the secular and military "defender of the faith." In the years of his reign (1519–1556), Charles would use his office in campaigns against his European rivals. In exchange, Pope Alexander VI (a Spaniard), secured for the Castilian Crown control over ecclesiastical matters as well as legal title over all newly discovered territories in the New World. Through this arrangement, Spanish governmental officials were able to establish almost total control over missionary programs, the internal governance of Church activities, as well as the selection of bishops, friars, priests, and lay officials. This allowed for an unprecedented accumulation of power among the Spanish nobility. In 1501, Pope Alexander had further strengthened the power of the monarchy over Catholic Church affairs by allowing the Crown to collect and administer religious taxes, or tithes, in the New World in exchange for the establishment of missions and churches in colonial territories. For the first time, the secular government of Spain had unparalleled powers over religious affairs in its American territories (Elliot 1969: 11).

Militarizing Jesus

Spanish notions of religious and descent-based hierarchies were developed long before contact with the Pueblos. Spain's conquest of the Americas coincided with the Reconquista of the southern boundaries of the Iberian Peninsula. Prior to this, Spain as a political entity had never existed. Spanish national identity was forged in an atmosphere of racial and religious segregation. During the period of the Moorish (Muslim) kingdoms, southern Spain was a model of religious tolerance and peaceful coexistence among Moors, Christians, and Jews. Following the Reconquista, Jews who refused conversion were expelled from their homes and their property confiscated. Those who remained and converted to Christianity were known as *conversos*; those who secretly practiced Judaism were known as *marranos*. All of these Spaniards would receive the brunt of Spanish discrimination and persecution during the Inquisitions of the sixteenth and seventeenth centuries. *Moriscos* (Muslim converts to Christianity) who remained in Spain fared no better. Forcible conversion had allowed both Jews and Moors to temporarily maintain their citizenship, but the continuation of alien cultural practices created tensions between "old" and "new" Christians. Eventually, descent-based, proto-racial classifications would emerge as the salient diacritical markers within Spanish society.

There were important economic incentives for ethnic subordination as well. The consolidation of political power by Spanish monarchs was accomplished through the dispossession and appropriation of Muslim and Jewish wealth (Kamen 1991: 9–57). The Reconquista institutionalized the legal basis for the use of military force and violence, forcible conversions, the enslavement of pagans, and the collection of tribute and labor among conquered peoples. The idea that Christianity, an ideology that explicitly rejects the accumulation of wealth and material possessions, that embraces pacifism and a radical egalitarian view of the rights of all peoples before God, could be both militarized and deployed as a means of acquiring wealth is one of the most fascinating (and jarring) philosophical ironies in human history. The notion that Christianity could be forcibly imposed on other groups was initiated by the philosopher Augustine in the fourth century. Faced with the defection of a rival sect of Christians in North Africa known as the Donatists, Augustine developed the first rational and legalized approach to the subject of violence and warfare initiated in the name of Christ. Using three words uttered by Jesus in the gospel of John, "Feed my Sheep" (John 21: 15; Luke 14: 16–24), Augustine interprets the statement of Jesus to Peter as a commandment. Later he was to use this passage to argue that all people, under the aegis of the Catholic (universal) Church, could be compelled (even by force) to come to the table of the Lord. These three words would serve as the legal basis and justification for the initiation of just wars, Holy Crusades, the Reconquista, the initiation of the African slave trade, and the conquest of the Americas (see Williams 1990: 1–118 and Phelan 1970: 5–28 for a thorough discussion).

During the Reconquista, the concept of a just war had allowed for the dispossession and legal enslavement of Moors who refused conversion or had resisted Spanish forces. The notion of resistance was critical in this legal construct. The use of arms in the defense of one's people or territory against a Christian person, so identified as an agent or representative of Christ, was interpreted as the rejection of Christ himself. This legal fiction allowed for the suspension of normal moral and ethical codes of "Christian" behavior and allowed the *agent* to use whatever means necessary in the subjugation of the resistor. In this manner, the Reconquista had provided the Castilian nobility with access to slave labor and land holdings of conquered peoples. Exported to the Americas, these legal mechanisms would figure prominently in establishing the Indian slave economy in New Spain's Northern Frontier.

The system of descent-based or "proto-racial" hierarchies was developed in this religious climate as well. As recent converts to Catholicism, conversos had held wealth as commercial entrepreneurs and bankers. Jews had been excluded from the Spanish nobility and relegated to certain professions. Usury, the buying and selling of time (which was thought to belong to God alone), was a profession specifically allocated to Jews. Their influence as bankers and businessmen during a period of rapid urbanization coincided with the beginning of the decline of the old feudal system—a system that had favored and sustained a Spanish noble class. Freed from the prohibitions placed on them by their religion, conversos began to compete with Christian nobles for positions of authority and offices within the Spanish state. This shift would generate a new system of classification; by the middle of the sixteenth century, the criteria for social advancement in the home country had shifted. Royal statutes excluded conversos and their descendents from high offices, and descent (a proto-racial system) replaced orthopraxy as a diacritical marker.

"Pure," "Old Christian" ancestry "thus became for the lower ranks of Spanish society the equivalent of noble ancestry for the upper ranks" (Elliot 1963: 223). As the policies of *limpieza de sangre* (purity of blood) spread, laws were passed in 1566 prohibiting the use of Arabic language, dress, and customs. In response to these cultural differences, in 1599, Archbishop Cisneros (Ximenes) of Toledo instituted a practice that would be adopted by colonial officials in New Spain. Mass baptisms and forcible conversions allowed for the partial integration of Moors and Jews within Spanish society, but statutes were written to limit the activities, mobility, and religious practices of stigmatized Jewish and morisco populations. Based on the performance of cultural practices specific to Judaism and Islam, Jews and Moors were continually targeted by Spanish officials during the years of the Inquisition. Initially, conversion to Christianity had allowed for a partial integration of Jews and Moors within the Spanish state. But as more people flooded into the status position of Christian, "Old Christians," those who were the descendants of Christian ancestors, found their existing status positions devalued. Eventually, they would assert new formulations of "authentic" and "pure" Christian blood. Descent from a Christian ancestor would determine access to power status within Spanish society. Foreshadowing the cycles of violence and resistance among stigmatized Native groups, Christian converts and moriscos protested against laws outlawing their traditions and staged large-scale, pan-ethnic rebellions against Spanish authorities in 1500 and 1570 (Elliott 1963: 51–53, 236–242).

This curious mixture of militant Christianity, Spanish nationalism, and descent-based racial hierarchies was exported to the Americas. Much of the cultural and historical symbolism associated with the conquests of Native peoples was infused with the discourses of the Reconquista. Santiago, the mythical brother of Christ whose pilgrimage site served as the spiritual nexus of Spanish military nationalism, was invoked in several sieges of the Pueblos. In naming the capital city of New Mexico Santa Fe (the name of the siege camp from which the final battle against the Moors was launched), the colonists projected onto the Pueblos the national mythology of a land many had never visited and that the Pueblos could only imagine. Reenactments of the conquest of the Moors were performed before throngs of perplexed Natives in Tenochtitlán, Tlaxcala, and Zacatecas (Harris 2000: 1–18). At San Juan among the Northern Rio Grande Pueblos, after the ceremonies of possession were performed, the bewildered Pueblos witnessed a similar performance: "the whole camp celebrated with a good sham battle between Moors and Christians, the latter on foot with harquebuses, the former on horseback with lances and shields" (Hammond and Rey 1953: 323). Conquest was highly symbolic, infused with the fervor of religious devotion and sustained through the embodiment of the colonist as an agent of Christ.

"Unjust, Scandalous, Irrational and Absurd": The *Requerimiento*, Silver, Slavery, and Rebellion on the Colonial Frontier

The use of Christianity as a tool of conquest attracted the attention of many of the finest philosophical minds and religious leaders in Spain. Bartolome de las Casas, the much-celebrated defender of the Indians and whose quote appears in this section's title, is credited as one of the most vocal critics of a militarized Jesus. His writings highlight the confusing struggle over the moral authority and voice of sixteenth-century Christianity. Although the rights of the Crown were theoretically subordinate to those of the Catholic Church, conquest was contingent on the conversion of Indios to Christianity. This relationship served as a legal and moral justification for the claims of the Crown on Indian land and labor. The rewards offered by forcible conquest offer clues to the motives and sensibilities of early explorers.

The *capitulation* was a contract between the Crown and the conquistadors that extended rights of property and governance in newly founded territories (Elliot 1969: 59). Through this relationship, the leaders of expeditions could acquire titles to land and labor as representatives of the king and Christ. Charters for settlements secured the rights of Spaniards to alienate property from Indian communities by force of arms. Codified in a document read before battles were launched against Native communities, the *Requerimiento* served to rationalize and routinize colonial bloodshed. Explained in a language many Natives had never heard, the Requerimiento outlined the cosmology, religion, and authority of the Spanish colonists and threatened death or enslavement to those who resisted. This peculiar legal document is explicit in explaining the mission of the Spanish state and justifying the use of violence in the subordination of Native peoples. It was also widely criticized by Spanish intellectuals and clerics. Still, the document demonstrates the degree to which the mythology of Christianity and violence were rationalized. Using the Petrinological thesis (where Jesus anoints Peter and by proxy his papal successors), Spanish jurists equated an assault on a Christian agent as grounds for the enaction of a total war (a war of fire and blood):

> With the help of God, we shall forcibly enter into your country and shall make war against you in all ways and manners that we can, and shall subject you to the yoke and obedience of the Church and their Highnesses: we shall take you, and your wives and your children, and shall make slaves of them, and as such shall sell and dispose of them as their Highnesses may command; and we shall take away your goods, and shall do all the harm and damage that we can, as to vassals who do not obey, and refuse to receive their lord, and resist and contradict him; and we protest that the deaths and losses which shall accrue from this are your fault, and not that of their Highnesses, or ours, nor of these cavaliers who come with us.[6]

Las Casas roundly criticized the Requerimiento in his *Historia de Las Indias,* condemning it as "*injusto, impío, escandaloso, irracional y absurdo*" (Las Casas, quoted in Hanke 1938: 33). Still, the Requerimiento reveals how social relationships (and distances) between Indians and Spaniards were conceptualized. The full text of the document (which must have taken several minutes to read) explains exactly what will happen to the Indians in the ensuing moments if they fail to comprehend and respond properly to the demands of the Spaniards:

> On the part of the King, Don Fernando, and of Doña Juana, his daughter, Queen of Castile and Leon, Subduers of the barbarous nations, we their servants notify and make known to you, as best we can, that the Lord our God, Living and Eternal, created the Heaven and the Earth, and one man and one woman, of whom you and I, and all of the men of the world, were and are descendents, and all those who come after us. But on account of the multitude which has sprung from this man and woman in the five thousand years since the world was created, it as necessary that some men should go one way and some another, and that they should be divided into many kingdoms and provinces, for in one alone they could not be sustained.
>
> Of all these nations, God our Lord gave charge to one man, called St. Peter, that he should be lords and Superior of all the men in the world, that all should obey him, that he should be head of the whole human race, wherever men should live and under whatever law, sect or belief they should be; and he gave him the world for his kingdom and jurisdiction.
>
> And he commanded him to place his seat in Rome, as the spot most fitting to rule the world from; but also he permitted him to have his seat in any other part of the world, and to judge and govern all Christians, Moors, Jews, Gentiles, and all other sects. This man was called Pope, as if to say, Admirable Great Father and Governor of men. The men who lived in that time obeyed St. Peter, and took him for Lord, King and Superior of the Universe; so also have they regarded the others who after him have been elected to the Pontificate, and so it has been continued even until now, and will continue until the end of the world.

This is a startling document. But it reveals the degree to which military tactics and the use of force were supported and even encouraged through an aggressively evangelical ideological movement—a militarized version of Jesus. Despite this outward show of zealous belief, the economic interests of colonists frequently clashed with the mandates of conversion sponsored by the Crown. Nowhere was this more evident than in the Northern Frontier. Just as the terms of difference developed in Spain were grounded in the historical interaction of Castilians Moors and other Spaniards, Indian people came to experience Spanish colonists through this strange and alien historical prism.

Spain may have been a "defender of the faith," but its opposition to Protestantism in England and Europe (Henry VIII was the king of

FIGURE 9. Indian Slaves in Spanish Silver Mines.

England from 1509 to 1547) was a costly position. The economic pressures exerted by Spain's wars against Protestantism fueled the drive for Indian land, labor, and silver. Frequently, these pressures forced the Crown to compromise its missionary efforts in favor of the more immediate needs of the home country. Imports of silver from the mines of New Spain increased from 85,000 kilos in 1530 to 300,000 in 1550 (see figure 9) With improved processing techniques and the discovery of mercury mines in Peru in 1570, import levels would reach 2.7 million kilos by 1590 (Elliot 1989: 19; Mackenney 1993: 57). Still, lacking an effective monetary policy (the health of the economy was simply tied to supplies of bullion) and a knowledge of economic theory, the influx of bullion created massive inflationary pressures. By flooding European markets with silver and gold, inflation in Spain increased 400 percent during the sixteenth century (Crown 1985: 166–167; Elliot 1989: 18–25; Kamen 1991: 88–99). Despite the windfalls provided by the silver mines at Potosí and Zacatecas, the Spanish government was forced to borrow heavily from European bankers to fund wars against the Protestants in the west and Turks in the east. By 1557, Charles V's wars of religion had bankrupted the Spanish government. Charles V's heir, Phillip II, conducted overseas acquisitions in Asia that added significant revenues, but the coffers of the Spanish Empire were exhausted in 1575 and 1596. Similar bankruptcies were declared during the reigns of Phillip III (1607), Phillip IV (1627, 1647, 1653), as well as Charles II (1680).[7] Increasingly, Spain turned to the resources of New Spain's Northern Frontier to supply the home country with cash.

Despite the official pronouncements of Spanish authorities banning Indian slavery, slave raiding was a frequent and familiar feature of the frontier economy. Native peoples were continually forcibly conscripted to work in the region's silver mines. The working conditions were often brutal. Historical accounts report that thousands upon thousands of Native peoples were worked to death, died in accidents, or were poisoned while working with the toxic agents of silver smelting (West 1949: 12). Faced with acute labor shortages and increased demands for bullion form Spain, frontier soldiers provoked attacks and raided slaves from Indian villages.

Citing the implied consent of the Requerimiento, the doctrine of *bellum justum* (just war) allowed for the legal enslavement of Indians who resisted Spanish authorities.[8] Reviled by secular and religious figures alike, the tactic would eventually contribute to a large number of Native revolts throughout the colonies (Powell 1953: 64). As the result of these practices, several Indigenous rebellions raged throughout mining towns in Zacatecas, Guanajuato, San Luis Potosí, and Pachuca from 1520 until well into the seventeenth century (Meyer and Sherman 1991: 161).

The Mixtón War erupted in 1541 when Francisco Vázquez de Coronado vacated Nueva Galicia in search of the Pueblos of Cibola (see figure 10). Stretching from Guadalajara to Culiacán, this regional insurrection lasting many years offers some striking parallels to the Pueblo Revolt of 1680. Tenamaxtli, the religious leader of the Mixtón Rebellion, urged his followers to "kill all Spaniards and burn their churches as the first step toward the return to old ways and old gods" (Miller 1985: 115). Unable to establish military control, Viceroy Mendoza was forced to enlist the aid of thirty thousand Aztec and Tlaxcallan allies. Still, control over the region was tenuous, and supply trains from the silver mining districts were continually disrupted. An era of pan-Indian rebellion had begun.

The Chichimeca Rebellion began in the 1550s and lasted for nearly forty years (Weber 1992: 212–227). Revolts among Tarahumara and

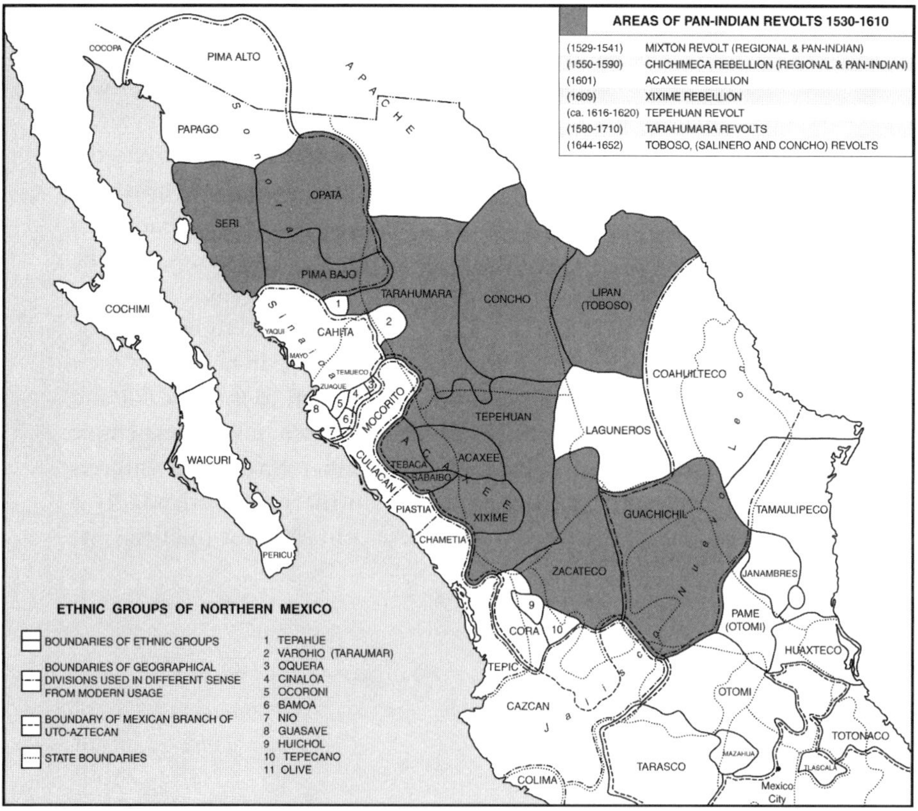

FIGURE 10. Areas of Pan-Indian Revolts 1530–1610.

Acaxee peoples in the sixteenth century were followed by the Xixime Rebellion in 1610 (Salmon 1991: 1–51). In 1616, the Tepehuan Rebellion added to the strains between colonists and Native peoples (Deeds 1998: 3–29). Unable to distinguish between secular and religious forms of social practice, Spanish colonists frequently interpreted alien ritual behaviors as acts of heresy. In an increasingly familiar pattern, Native religious practices were outlawed, and religious leaders were arrested. As the suppression of traditions increased, Native peoples retaliated by defiling Spanish missions and attacking church officials. As the demands for a stable source of Indian labor increased, so did the frequency of slave raiding.

The cycle of violence and retribution was a continual concern of colonial officials. Responding to the crisis, the fourth viceroy of New

Spain, Don Martín Enriquez de Alemanza (1568–1580), issued an order for a war of fire and blood, demanding the total elimination of all rebellious natives on the Northern Frontier (Powell 1953: 105). Still, there was little agreement among colonial officials regarding these practices. In a confidential letter to King Phillip II, Juan Buatista de Orozco, the military leader appointed to quell the rebellions in the northern Borderlands, stated that the wars against Native peoples had been waged in an attempt to secure Indian slaves. Buatista writes, "In the beginning, this was the easiest solution, but now this right to spoils has become the principle goal of the soldiers and continues to be the cause of great excesses. The capture of free Indians to use in paying off their debts to the merchants will only cause great harm to us" (Naylor and Polzer 1986: 46). Faced with few economic opportunities, expensive supplies, and deflated silver prices on the frontier, Spanish adventurers looked for new opportunities along the northern Borderlands. It is within this context that the Pueblos were contacted for the first time in 1540.

CHAPTER 5

Abandonment as Social Strategy

Colonial Violence and the Pueblo Response

At the 2008 Society for Historical Archaeology Conference (SHA) in Albuquerque, New Mexico, a group of historians and archaeologists gathered to discuss some new discoveries and developments in the scholarship of the Coronado expedition and Spanish entradas. Sponsored by historical archaeologists Clay Mathers and Charles Haecker and attended by Coronado historian Richard Flint, the symposium was the first of its kind in a number of years. Packed with scholars and Coronado enthusiasts, the session focused on the use of an old and much-maligned tool of amateur archaeology: the metal detector. Sifting through a number of sites along the now largely urban Albuquerque course of the Rio Grande, archaeologists were startled by the presence of significant concentrations of small metal objects—many too small to be recognized or collected in traditional quarter- and one-eighth-inch screens—and all so caked with corrosion that they were totally indistinguishable from the small dirt clods that frequently clog sifting screens.

Usually, these objects would pass through the normal filters of evidence and be lost forever. The researchers had revisited previously surveyed sites at Zuni (Haikuh) as well as a Tiwa site (Piedras Marcadas) and returned with their findings. The discoveries were startling. Researchers were able to map out the precise locations within a village where metal shot (fired from large front loading harquebuses) or bolt heads from crossbows accumulated. For the first time in centuries, the tactics and patterns of attack and defense could be reconstructed.

This was new information. And it reignited interest at the conference in the subject of colonial violence and its effects on Pueblo peoples. There still exists a strong reaction against discussions of violence among many Borderlands historians. And the most vocal critics are, not surprisingly, those who interpret abandoned contact period sites as evidence of epidemics.

The differences between these two interpretations are most evident in the logical conclusions that each one affords. In the disease model, abandoned sites are interpreted as the skeletons of a dead culture. From this interpretation, the mythology of the ever-vanishing primitive evolves. Every site, by virtue of its abandonment, can be interpreted as resulting from some form of social, technological, or environmental (in this case, epidemiological) failure. This is largely the result of the nature of archaeological evidence (archaeologists don't usually study places where Indians still live) and of the professional and disciplinary boundaries that made discussions with descendant communities and actual live Indians passé. The other interpretation is one that envisions archaeological sites the way Native peoples see them: as part of a living cosmological and historical landscape—landscape that is still inhabited by contemporary Native peoples. In one, the site is a corpse; in the other, the site is like the shell of a hermit crab—the outside is shed, and the living thing inside, the people who inhabited the site, move on to new locations.

Change is a continual feature of every human society, but in the case of Native prehistory, change is frequently rendered as some kind of catastrophe. This is *essentialism*—the belief that the true Native people lived in the past, somehow uncontaminated by modernity and that when contacted by Europeans, they lacked the natural ability to defend themselves and insulate their cultures from outside influences. It assumes that Native peoples had no knowledge of otherness, that the only reason they lived the way they did was because they had not yet been contacted by peoples they considered "different." It mirrors several dimensions of the "Indian as savage innocent" trope that is commonly projected on Indians from the outside; it assumes a simplicity of thought, an inability to reason or think creatively (the idea that Spaniards were gods), and a "natural" or primal state of being untainted by the whole range of human experience. It is what I like to call the "ooga-booga syndrome" and appears in almost every film dealing with contact. Indians are depicted as horrified and fascinated by the presence of unfamiliar clothing and light skin and run in horror from their sight shouting, "Ooga-booga!" Europeans are depicted as emotionally centered and intellectually unfazed by otherness. There is no sense of naiveté or childlike simplicity.

The disease model follows many of these themes: An innocent and untainted population is invaded by pathogens that they had no experience with or resistance to. The truth is that diseases such as syphilis and some kinds of tuberculosis are present in precontact populations. Despite the fact that Europe was invaded by diseases that came from Central and East Asia, I have never heard a discussion about the "vanishing European" or "acculturation among the Spaniards." This is a mythology that archaeologists have helped create and that continues to alienate and irritate Native Americans. There are precious few narratives, archaeological or historical, that explain the presence of Indian peoples. Beginning with the assumption that the Pueblos are still here, that events such as the corn dance still have meaning, we can account for that presence and reverse some of the mythology surrounding the terminal narratives of contact. This requires a different set of analytical tools and a rereading of documentary histories for clues as to how survival and persistence might be explained.

This chapter explores the correlation between social violence, abandonment, and mobility during the entradas of the sixteenth century and the seventeenth-century early colonial periods in New Mexico (see table 5.1 and figure 11). Of particular interest to archaeologists are the behaviors, responses, and reactions of the Pueblos to the Spanish colonists. Using primary sources, I document the use of violence during the entradas as a means of establishing superior and subordinate relationships between colonists and Pueblos as well as the collective responses of the Pueblos to these actions. While armed resistance was a common feature of many of these interactions, the Pueblos also attempted to defend themselves and maintain social distances through migration, regional abandonment, and mobility. Though generally unexplored by archaeologists or historians, these "refugee movements" are a characteristics shared by stigmatized populations in colonial contexts. More importantly, the documentation of these movements provides an alternative explanation to traditional disease-motivated, depopulation theories and argues that violence, not disease, led to the abandonment of many contact period sites. It also provides the context in which the Pueblo Revolt (and its predecessors in northern Mexico) can be understood. Violence and subjugation in the Pueblo world led to attributes encountered by visitors to this day—the aggressive defense of historical, religious, and cultural knowledge from outsiders.

TABLE 5.1. New Mexican Entradas, Viceroys, and Kings

Expeditions	Years	Viceroy	King
Cabeza de Vaca	1528–1536	Guzman (1529–1533)	Charles V (1516–1558)
		Mendoza (1535–1550)	
Fray Marcos de Niza/Esteban	1539	Mendoza (1535–1550)	Charles V (1516–1558)
Coronado	1540–1542	Mendoza (1535–1550)	Charles V (1516–1558)
		Luis de Velasco Sr. (1550–1564)	Phillip II (1556–1598)
		Gaston de Peralta (1566–1568)	Phillip II (1556–1598)
		Martin Enriquezde Alamanza (1568–1580)	Phillip II (1556–1598)
Fr. Augustin Rodriguez/Capt. Chamuscado	1581–1582	Lorenzo Suarez de Mendoza (1580–1583)	Phillip II (1556–1598)
Espejo	1582–1583	Lorenzo Suarez de Mendoza (1580–1583)	Phillip II (1556–1598)
		Pedro Moya de Contreras (1584–1585)	Phillip II (1556–1598)
Castano de Sosa	1590–1591	Alvaro Manrique de Zuniga (1585–1590)	Phillip II (1556–1598)
Morlete	1591	Luis de Velasco Jr. (1590–1595)	Phillip II (1556–1598)
Leyva de Bonilla/Humana	1593	Luis de Velasco Jr. (1590–1595)	Phillip II (1556–1598)
Onate	1598	Gaspar de Zuniga y Acevedo (1595–1603)	Phillip II (1556–1598)
Juan Martinez de Montoya	1600	Gaspar de Zuniga y Acevedo (1595–1603)	Phillip III (1598–1621)
			Phillip IV (1621–1665)
1680 Pueblo Revolt/Otermin Gov.	1680		Charles II (1665–1700)
Vargas Reconquista	1692		Charles II (1665–1700)
1696 Pueblo Revolt/Vargas Gov.	1696		Charles II (1665–1700)

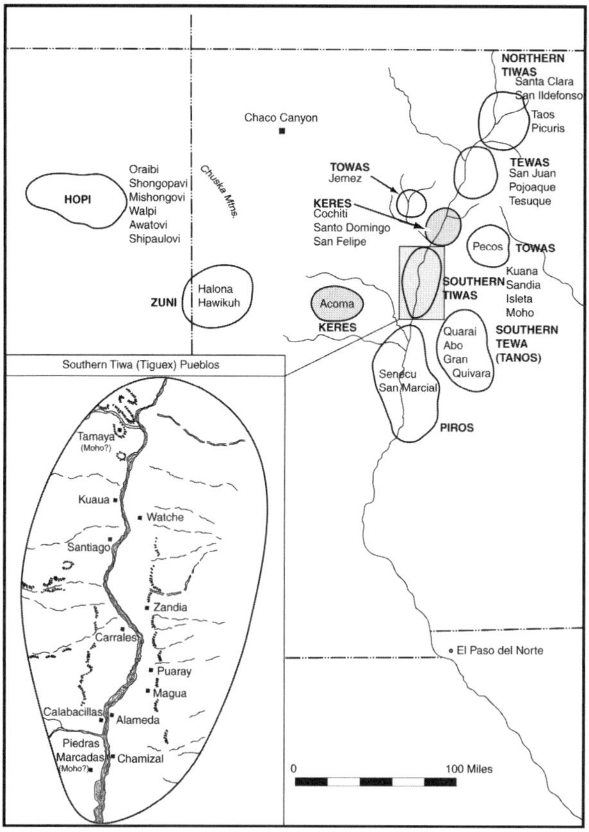

FIGURE 11. Historic Pueblos With Inset: Southern Tiwa Pueblos Mentioned in Coronado Entrada.

Like many peoples under siege, these experiences can animate and strengthen human communities and actually initiate powerful countervailing movements that defy typical notions of Indian passivity and the supposed "inevitability" of acculturation. This approach has applications in almost every other Indigenous/colonial arena, but it also should reformulate ideas about the reactions of any subject people when faced with either violence or the aggressive expansion of empires.

The subject of abandonment raises an equally important related question: If mobility was a more common event than has previously been recognized, where did the Pueblos move to? Although many of the abandonments in the seventeenth century are attributed to Apache and Navajo raids (particularly in the Piro and Tompiro Pueblos), an equally strong case may be made for the influence of slave raiding by Spaniards and the establishment of alliances with the more mobile communities of Apache and Navajo allies.[1] It is not surprising that while historical estimates of Pueblo population are reported to decline (a phenomenon discussed later), estimates of Apache and Navajo populations explode. Compounding the problem is the fact that most historic Puebloan villages and refugee sites have not been systematically studied. Many of the remote, large sites in the Jemez Mountains show evidence of continual or intermittent occupation during the sixteenth and seventeenth centuries. Given the centrality of these communities during the Revolt of 1680, the Jemez Pueblos serve as a regional case study for the subject of regional abandonments and reoccupations.

The collective experiences of the Pueblos during the entradas and early colonial period also determined the relationships among Native communities and between the Pueblos and colonists in the seventeenth century. The efforts of missionaries and colonists to forcefully convert the Pueblos, eradicate traditional religions, and extract tribute and labor from them helped generate the nascent pan-Puebloan identity. While diseases played a significant role in reducing Native populations, the demands of colonists to support both Pueblos and colonists severely compromised the survival strategies of the Pueblos. Excessive tribute levels and the appropriation of land, water, and labor by secular, private, and church officials pushed the Pueblos to their limits. Ultimately, the commonly shared experiences of ascription, stigmatization, and subordination provided the Pueblos with a collective history and a pan-Indian self-consciousness that helped galvanize the Pueblos during the Revolt of 1680.

While the objectives of seventeenth-century colonization represented a more balanced integration of secular and religious interests, the entradas of the sixteenth century were motivated by a desire to secure access to the resources that fueled the expansion of the Spanish Empire. Between 1540 and 1598, the Pueblos encountered at least seven successive waves of disappointed plunderers, adventurers, and profiteers seeking wealth and fortune. Despite the promises and legends of large tribute-rich northern cities inspired by Cabeza de Vaca's journey from Florida to Mexico City in 1538 and Fray Marcos de Niza's foray to the Northern Frontier in 1539, the Pueblos lacked not only the mineral wealth that the Spaniards demanded but were incapable of providing the levels of tribute required to sustain both the colonists and themselves

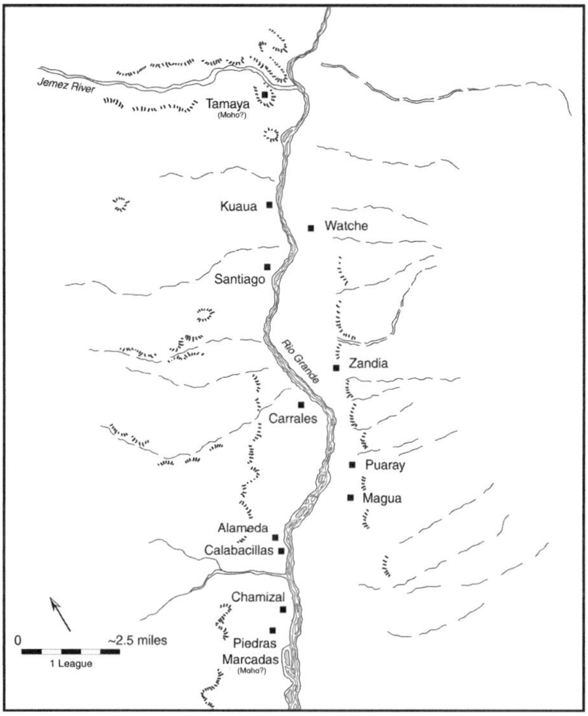

FIGURE 12. Southern Tiwa (Tiguex) Pueblos.

during these expeditions. From Coronado's failed expedition to find the fabled cities of Cibola to the eventual settlement of Juan de Oñate at San Juan in 1598, encounters between the colonists and Pueblos were characterized by a series of violent confrontations. These confrontations generated deep resentments among Puebloan communities and their neighbors and established the terms and character of the interactions between the two groups throughout the seventeenth century.

Coronado staged violent attacks at several pueblos including Zuni, Acoma, and Hopi (see figure 12). More damaging were extended sieges of thirteen of the fifteen Tiwa communities during the winters of 1540–1542. Both Arenal and Moho were the sites of particularly devastating attacks (Hammond and Rey 194: 226–230, 333–358). Coronado's demands for food, clothing, and shelter as well as his capture of religious leaders at Pecos generated ill feeling toward the colonists. For six months, Coronado's men camped at Tiwa villages and staged expeditions to look for silver, lead, and mercury. As their demands increased, tensions flared between the groups. Following the legal means of dispatching idolaters, two hundred Tiwa men were burned at the stake (223–234, 330–336). These attacks were followed by a sustained and costly siege at the Pueblo of Moho. According to the official testimony provided at Coronado's Trial, more than four hundred natives had been slain in the attacks (337–393). Ultimately, both Coronado and his men were charged by the Spanish governors for crimes against the Indians. The expedition and the lengthy trial afterward were to bankrupt Coronado, who died in Mexico City in 1544.

For forty years, no documented contacts occurred between the colonists and Pueblos. Slave raiding and other Indigenous rebellions continued throughout the period. Coronado's journey was followed by the Chamuscado-Rodriguez, Espejo, and Castaño de Sosa expeditions of 1580–1581. In each case, a similar cycle of violence was touched off as demands for tribute, servants, and women pushed the Pueblos to their limits. In regions that were especially hard-hit by colonial raids, the colonists encountered and recorded the presence of villages that had been completely abandoned. When the Spaniards asked where the inhabitants had gone, they were told that the people had fled into the surrounding hills (Hammond and Rey 1929: 71–79).

Finally, in 1598, Juan de Oñate established a permanent settlement at the Pueblo of San Juan after reconquering the Pueblos once again in the name of Phillip II. Once again Acoma was besieged by the colonists. After one of Oñate's nephews was killed in an attack on the pueblo, Oñate ordered a war of fire and blood against the Acomitas. The Spaniards claim to have killed eight hundred people in the village (Hammond and Rey 1953: 427). The remaining five hundred captives were divided among Oñate's party; men over the age of twenty-five had a foot severed and served twenty years servitude, and women over twelve were to serve twenty years. The female children were given to the friars, and the boys were given to Oñate's captain, Vicente de Saldívar (Hammond and Rey 1953: 21–23). The prisoners were then transported with the soldiers to the colonial capitol at San Juan. The location of the capitol is most significant. The generation of the pan-Puebloan movement depended in large part on the common experiences among the various communities of the Puebloan world. Po'pay was born and reached prominence as a spiritual leader at San Juan. As a resident of the pueblo, historical memories of the punishment and enslavement of the Acomitas must have aided Po'pay in his cause.

Language and Communication as Social Boundary

Of great importance in determining the causes and circumstances surrounding many of the early violent confrontations between Spanish explorers and the Pueblos is the subject of communication. Notions of social distance are largely based on historical relations between various ethnic groups. Spanish conquistadors were at a distinct disadvantage, linguistically and socially, in communicating their interests with previously unknown groups along the frontier. The Pueblos, in contrast, had individual historical experiences, protocols for communication, and trade relations with other ethnic groups in the surrounding areas. The Pueblos are represented by at least five distinct language groups: Zuni, Hopi, Keres, Tiwa, Towa, Tewa, and Piro (see figure 13). These are further segregated by northern and southern variants for Tiwa, Tewa, Keres, and Piro. A well-traveled Pueblo would have to know seven distinct languages (not to mention Navajo or Apache). It is difficult to believe that Native guides brought north from the Valley of Mexico would have been able to fluently converse clearly within this densely packed valley of foreigners. Linguistic barriers allowed the Pueblos to conceal their intentions, speak freely, and communicate with one another on levels that the colonists could not comprehend. It seems likely that early contacts with Spanish colonists generated a collective notion that miscommunication, violence, and the violation of normal protocols could be expected during these exchanges.

Given the presence of long-distance communication and trade networks between the Pueblos and their southern neighbors, Puebloan people were likely well aware of the threats posed by the entrance of the Spanish into New Mexico. Although historian Carroll Riley claims that the Pueblos were "not very well informed as to the real purpose of the Spanish Army" (1971: 303), the long-term historical relationships between various trading partners in the immediate region (as well as northern Mexico) provided the Pueblos with a distinct social and informational advantage over the Spaniards.[2] Evidence for Pueblo trade with their more powerful and numerous neighbors in the Valley of Mexico is authenticated by the testimony of Coronado entrada member and chronicler Pedro de Castañeda. In preparing for the journey north, Castañeda enlisted the aid of a Native familiar with the trade routes from the Valley of Mexico to the Pueblos.[3]

Drawing from oral histories at the pueblo, Zuni anthropologist Edmund Ladd suggests that communication networks between the

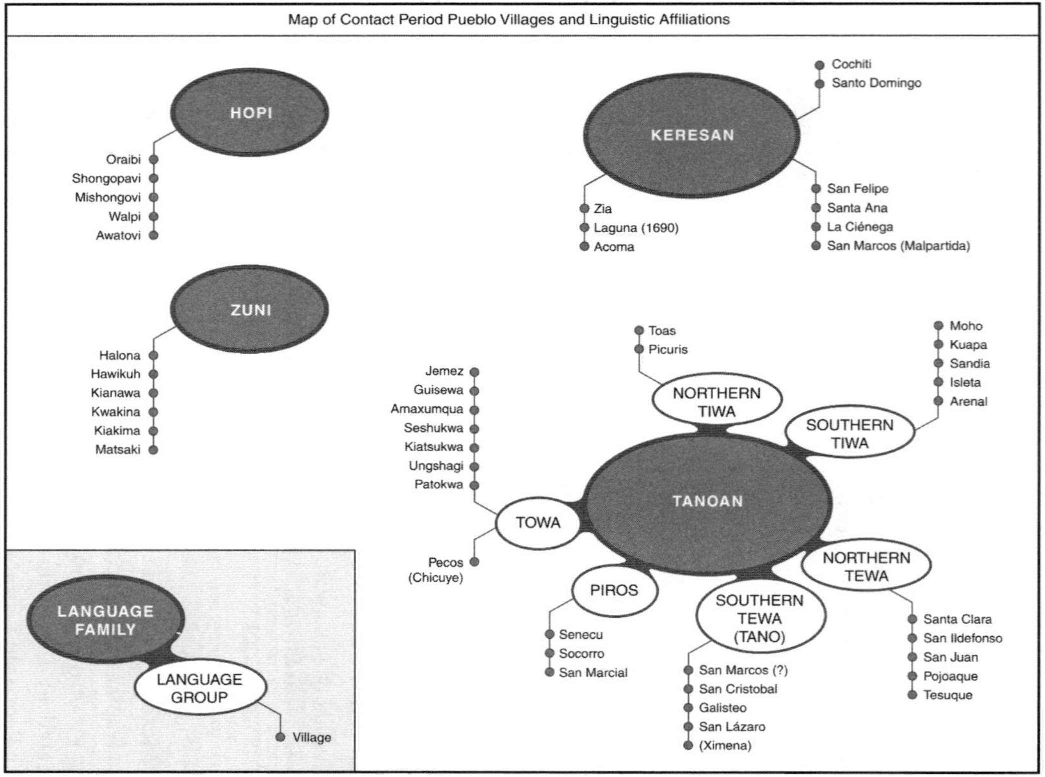

FIGURE 13. Linguistic Affiliations: Pueblos. The Pueblos are among the most linguistically diverse concentration of Native settlements in the Americas. Each language family is as distinct from one another as Chinese might be to German. It is highly unlikely that Spanish colonists or missionaries were able to learn and converse with the Pueblos on the same level as the Pueblos, who only had to learn one (Spanish) language. Instead, they relied on Indian translators who held tremendous power within the colonial system. These Indians (or *coyotes*) were highly influential in organizing rebellions.

Zunis and their southern neighbors informed the Pueblos of the behavior, weaponry, and slave-raiding activities of the Spanish years before contact was made (1997: 225–234). This claim is supported by Spanish documents. According to Coronado, fifty years prior to his arrival (ca. 1490), the Pueblos had been informed that "a people such as we would come, and from the direction we have come and that the whole country would be conquered" (Hammond 1940: 175). Coronado also mentions having been monitored through a system of fire signals that "had warned of our coming and where we had arrived" (Hammond 1940: 167).

In this sense, the ascriptive process was dialectical. As evidenced by the responses of the Pueblos to the Spaniards' behavior, the ascriptive process (finding out what kind of people these are) is revealed by the collective behaviors of the Pueblos. Incidents of village abandonment, the segregation of women and children, and the repeated pattern of armed resistance and appeasement all suggest that the Pueblos had formulated (through communication networks) a collective portrait of how the Spaniards could be expected to behave. Most instructive is the correlation between social violence and mobility in the Tiwa region. The actions of the Tiwas foreshadow the most drastic and costly tactic employed by the Pueblos. In an effort to maintain social distances between themselves and the colonists, the Tiwas abandoned their traditional homelands, spiritual landscapes, resources, and ancestral villages; in some cases, these movements were permanent. The following section outlines the movements, conflicts, and tactics of resistance offered by the Pueblos during six of the entradas of the 1500s. The final section explains how tribute demands, forcible conversion, and the suppression of Native cultural practices led to a series of minor rebellions. Collectively, these interactions helped generate the regional pan-Puebloan movement of the Revolt.

The Coronado Entrada: 1540–1542

The behavior of Coronado's men during the 1540 expedition would signal to the Pueblos the presence of an unpredictable, rash, and desperate foe. The sieges of Arenal and Moho during the Tiguex War were incidents that helped define the character of later interactions between the groups. Coronado was serving as a *regidor* (similar to a city councilor) in Mexico City when Álvar Núñez Cabeza de Vaca and an African slave known as Estébanico returned from an epic voyage through the southern and southwestern interior of North America. Initially shipwrecked near present-day Tampa, Florida, in 1528, four survivors journeyed for eight years until the finally reached Mexico City in 1536. Their fantastic and wildly exaggerated descriptions of the wealthy communities farther north were heard by Viceroy Mendoza, Hernán Cortés, Coronado, and Hernando de Soto (Bolton 1949: 40). The viceroy purchased Estéban and quickly dispatched Fray Marcos de Niza in order to investigate de Vaca's reports. The specifics of his journey have been recorded elsewhere (Bandelier 1893; Bolton 1915, 1949; Hammond 1940; Hallenbeck 1987) and, like de Vaca's account, is more remarkable for its exaggerations, hyperbole, and the rumors of unconquered wealth it inspired than anything else.

In his deposition to the viceroy, de Niza stated that he had

> stood at the outskirts of Cibola (Zuni) and that the while the city was larger than Mexico City, it was the smallest of the seven cities of Cibola and contained more than 20,000 houses. The people are almost white. They wear clothes and sleep in beds. . . . They possess many emeralds and other jewels, although they prize none so much as the turquoises. . . . As they possess no other metals, they use vessels of gold and silver, of which they make greater use, and there is greater abundance of it than in Peru. They buy it with the turquoises at the province of the Pintados where it is said that the mines are found in great abundance. (report of Fray Marcos de Niza to Viceroy Mendoza, September 2, 1593, in Hammond and Rey 1940: 79)

Viceroy Mendoza, a personal friend of Coronado, appointed the twenty-eight-year-old *peninsulare* nobleman to serve as the replacement for Governor Nuño de Guzmán in the rapidly expanding northern mining province of Nueva Galicia. In the previous year, Nuño de Guzmán had been recalled to Spain and imprisoned for illegally enslaving Indians in the mines of the region and inciting the first Chichimeca rebellion in 1538 (Salmon 1991: 1–29; Naylor and Polzer 1986: 150). The familiar Spanish tactic of provocation, violent raids of Native villages, capture, and forcible labor had led to large-scale abandonments and outright rebellion in the province. Coronado's attentions were focused on securing a more stable population of Native tributaries and laborers. In his letters to King Charles V (Hammond 1940: 35–41, 45–49), Coronado explains the labor shortages as resulting from an illicit and widespread slave trade between the province and Mexico City. "Some of those who deal in Indians buy and sell unbranded and free Indians as slaves. . . . Because of the scarcity of slaves, the mines are being worked by Indian slaves and free Indians" (38).

While serving as the governor of Nueva Galicia, Coronado had accompanied Fray Marcos and Estéban for part of the journey north in their attempt to locate Cibola. Coronado was known and rewarded for his brutality in dealing with the pan-Indian Mixton Rebellion. After a relentless suppression in which hundreds of Indians were killed, Coronado apprehended a leader identified as Ayapin; and in his letter to King Charles, he describes how he was drawn and quartered. In the same letter, Coronado explains in great detail the manner in which slave raiding and Indian slavery have served as the engine of the economy: "The manner in which residents support themselves is that in the gold mines, most of the towns give their *encomenderos* [recipients of tribute] Indians to get out the gold for them. Some of those who deal in Indians buy and sell unbranded and free Indians as slaves." He also informed King Chares V of Marcos de Niza's findings in the northern Borderlands (Hammond 1940: 37, 45–49).

By January 1540, Coronado was appointed as leader of an expedition to the Pueblos. Accompanied by Marcos de Niza, his force consisted of 230 horsemen; 62 foot soldiers; 1,000 horses; 600 mules; and 1,300 Aztec, Tlaxcallan, and Tarascan Indian "allies and servants from Nueva Galicia" (Hammond 1940: 87; Riley 1997: 6). Frustrated by the slow pace of the expedition, in April 1540 Coronado led a smaller force of 120 men north to find the Pueblos. Fray Marcos's distortions are legendary. A map drawn in the 1560s (figure 14) shows Zuni (Cibola) as one of a group of seven large cities. Niza's description of Zuni as comparable in size to Tenochtitlan (Mexico City in the bottom right corner) is evidenced by the distortions of the map. Zuni is rendered as almost as large as the island of Saipan (Japan). Coronado's fury was matched by the ferocity of his attacks on the Pueblos. Convinced that larger cities lay inland, the Spaniards and their Indian allies captured, murdered, and tortured Pueblo leaders in an attempt to learn where the real wealth was hidden.

On July 7, Coronado reached the Zuni pueblo of Hawikuh. After reading the residents the Requerimiento, Coronado's forces, lacking food and supplies, were immediately attacked by Zuni warriors and staged a series of counterattacks. During the day-long battle, Coronado was incapacitated, receiving injuries to his head and foot from arrows and stones.[4] In a letter to the viceroy, he expressed his disaffection with Fray Marcos de Niza—"I can assure you that he has not told the truth in a single thing that he said, but everything is the opposite of what he related" (170)—and documents the removal of women and children

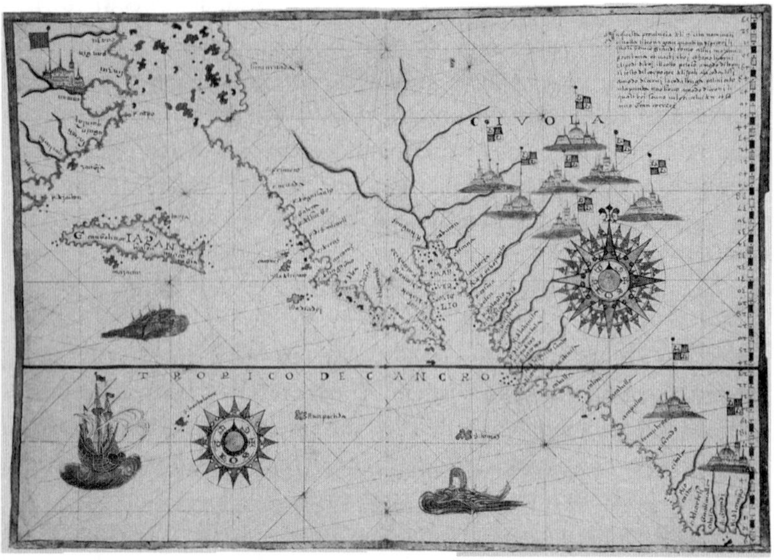

FIGURE 14. Sixteenth-Century Map of Cibola, Mexico City, and Japan. This map, published in the 1570s, shows the influence of Fray Marcos de Niza's distortions of Zuni. Fray Marcos described Cibola (Zuni) as the smallest of the seven cities and as large *(grande)* as Tenochtitlan (Mexico City drawn in the bottom right corner). Whatever *grande* referred to, it is clear by the relative sizes of Cibola and Saipan (which the Spanish thought of as Japan).

from the pueblo during the early stages of the encounter: "I did not find any women here nor any men under fifteen or over sixty. . . . I am unable to give your lordship any information about the dress of the women, because the Indians keep them guarded so carefully that I have not seen any, except two old ones" (170, 177). Three days later, Coronado reports that the entire pueblo had been evacuated:

> Suddenly, the next day, they packed up their goods and property, their women and children, and fled to the hills, leaving their towns deserted, with only some few remaining in them. . . . Seeing this, I went to the town which was larger than this. I found only a few natives there. . . . Later an old man, who said he was their lord, came. . . . He said he would come to see me with the rest of the chiefs of the country. Three days later, I order to arrange the relations which should exist between us. . . . They agreed to come down from their strongholds and return to their houses with their wives and children . . . and become Christians . . . but they still remain in their strongholds, with their wives and all of their property. (175)

The initial confrontation at Zuni reveals several clues as to the disposition of the Pueblos toward the Spanish. Despite a common misconception that Native peoples uniformly regarded the Spaniards and their horses as immortals, Coronado was singled out for attack at Hawikuh (Hammond 1940: 208); at Hopi, Coronado's men and horses were similarly targeted (214). From Zuni, Coronado dispatched parties to Hopi and Acoma and was visited by representatives from Pecos, including a Towa leader referred to as "Bigotes." While missionization was an eventual goal of the party, Coronado reveals that his chief objective was the establishment of a mining district in the region. "As far as I can judge, there does not appear to be any hope of getting gold or silver, but I trust God that, if there is any, we shall get our share of it, and it shall not escape us through any lack of diligence in the search" (177).

Learning of the presence of larger communities to the southeast (the Rio Grande Pueblos), Coronado sent his representatives (with Bigotes) to the region to prepare quarters for the remainder of his party. Forcing the residents of the pueblos to evacuate a pueblo named Alconfoor (Santiago, or LA 326),[5] Coronado spent the next two winters inhabiting the plundered villages of the Tiwas.[6] The strain of supporting nearly a thousand colonists was compounded by the fact that the Tiwas missed two seasons of planting and harvesting. Coronado's commander López de Cárdenas was unrepentant at his trial, stating that the residents gave up their homes willingly. "The Indians at the pueblo of Coofor [Alconfoor,] seeing that the said Don García López [Cárdenas] wanted to put up houses, told him not to build them as they would vacate their pueblo for him and go to other neighboring ones where they had friends and relatives" (Hammond 1940: 329).

While at Alconfoor, associations between the Pueblos and the Spaniards worsened. After being told that gold was available in a region to the east, Coronado arrested, interrogated, and tortured both Bigotes and a man from Pecos referred to as Cacique (Hammond and Rey 1940: 330). Resentments grew as Coronado's men forced the Tiwas to provide food, clothing, and shelter for the men.[7] Relations between the groups reached a breaking point when a Tiwa man from the pueblo of Arenal accused a member of Coronado's party of raping his wife (225).[8]

In retaliation, the villagers from the pueblo where the incident occurred seized and killed thirty to fifty Spanish horses (331). Two of Coronado's captains, Don García López de Cárdenas and Alvarado, returned to the pueblo, attacking the Tiwas and burning the village.

After a lengthy battle in which some of the Tiwas were captured and surrendered, those Tiwas who had not capitulated were hunted down and killed as they fled the settlement. In a tactic used repeatedly in attacks on Indians in the frontier, Cárdenas ordered the two hundred surrendered captives burned at the stake. Having now formally resisted the agents of Christ, this was in keeping with the legalized version of heretical execution. Cárdenas required that Bigotes and Cacique would be forced to witness the carnage, "in order to impose a punishment that would intimidate the others" (226).[9]

Coronado's men were not done. Out of provisions and eager to intimidate the Pueblos into meeting his needs and demands, Cárdenas found himself on the defensive and was attacked as he approached the village of Moho. Coronado personally led the attack, routing a party of Pueblo leaders who had likely emerged unarmed, for negotiations. Here for fifty days the Tiwa held out against the Spanish and in extreme thirst attempted to dig a well in the central plaza. The villagers released a large group of women and children (whose fate was unknown). In the last of several attacks against the pueblo, the people of Moho attempted to flee the village in the night. Pueblo men encircled women and children and attempted to cross the frozen landscape. They were attacked while crossing, "They fell back to the river, which was high and extremely cold, and as the men from the camp [soldiers] quickly rushed to attack, few of the enemy escaped death or injury. In the morning the army crossed the river and found many wounded Indians who had collapsed because of the intense cold" (230, 290, 334, 395). During the two-month siege, Coronado requested the assistance of the people of Pecos, but the Pecos "refused to grant him the favor he was asking them, excusing themselves by saying that they were busy with their plantings" (331).[10]

At Arenal, a similar scene ensued. After an extended siege lasting two weeks, the remaining Indians were apprehended while attempting to leave the pueblo at nightfall.

> The natives were leaving and going into the country. . . . The soldiers left the ambush and went to the pueblo and, seeing the Indians in flight, they pursued and killed many of them. As news of this was sent back, some of the men of the army came and sacked the pueblo. They apprehended all of the people they found there, comprising about one hundred women and children. This siege was completed by the end of March, 1542. (230)

The fate of the women and children is unknown.

In retaliation for what he considered the Tiwas' resistance, Coronado ordered his men to burn the ten remaining Tiwa villages (332–333). During Coronado's two-year stay in the Upper Rio Grande, the Tiwa villages remained completely vacant; "the twelve pueblos of Tiguex were never resettled as long as the army remained in the region, no matter what assurances were given them" (233–234).

From here, Coronado's men visited Zia and an unnamed Queres settlement where they found the villagers abandoning the pueblo in advance of their party. "At the first Pueblo, which must have contained one hundred residents, the people ran away, not daring to wait for our men. The latter [Coronado's soldiers] ran to intercept them and brought them back, fully protected to their homes. From here the Spaniards sent word to the other pueblos in order to restore their confidence. Thus the whole region was gradually reassured" (233).

Castañeda also reveals how social boundaries were maintained between the Pueblos and the "other" group of foreigners in the region: "The Teyas . . . were known by the people of the towns as their friends . . . [and] often go to the latter's pueblos to spend the winter, finding shelter under the eaves, as the inhabitants do not dare to allow them inside. . . . At night the visitors do not stay in the pueblo, but outside under the eaves" (258).[11] Later, Castañeda's report documents (among others) the presence of twelve Tigua (Tiwa) pueblos, seven Quirix (Keres) pueblos, seven at Hemes (Jemez), and eight Tutahaco (most likely Piro) villages to the southeast.[12] Refuting de Niza's highly inflated estimates of Puebloan population figures, Castañeda estimates that in the entire region there were a total of sixty-six pueblos "containing 20,000 men" (259).

By late fall, word had apparently spread regarding the activities of the colonists. At Pecos, Coronado's party was refused entry and attacked. Returning south to the Tiwa-Piro region, the colonists found that the people of the province had "gone back to settle, but had again fled again on account of fear" (243). Short on supplies, Coronado's men visited seven Jemez villages where they received provisions. But at San Juan, the home village of Po'pay, the Spaniards encountered an increasingly frequent occurrence. Wary of the activities of the colonists, the Pueblos increasingly exhibited the behavior of abandoning their homes prior to the arrival of the Spanish and heading into the mountains: "Those of *Yunque-Yunque* abandoned two very beautiful pueblos which were on opposite sides of the river, while the army was establishing camp, and went to the sierra where they had four very strong pueblos which could not be reached by the horses because of the craggy land" (244).

With the Tiwa Province evacuated, Coronado's party spent the remainder of the winter at Alconfoor (243). Disappointed by the absence of silver or gold deposits and worried that his own party would desert, Coronado made arrangements to abandon the region and return to Mexico City.[13] Meanwhile, reports reached the expedition of a Native rebellion in Nueva Galicia, and by April 1542, the Spaniards departed the region. In Nueva Galicia, Coronado learned that he and his captains had been charged with crimes against the Pueblos. Found guilty, he returned to Mexico City and successfully appealed the charges.[14] Cárdenas was similarly charged, found guilty, and sentenced to exile in Africa (Hammond and Rey 1940: 341).

The records of testimony at Cárdenas's trial depict a harrowing journey and an unrepentant defense of the attacks. The trial contradicts Coronado's account as to the causes of the rebellion and the use of violence as a means of subordinating the Pueblos. When asked if he knew why the Tiwas had rebelled, Cárdenas stated that "no cause had been given by the Spaniards" and that "they did it of their own accord" (349). At Arenal, Cárdenas corroborates others' descriptions of Coronado's tactics. He states that Coronado had ordered Bigotes and his party to witness the burning of the two hundred Tiwas, "so that they might see the power with which the Pueblo was taken, and the punishment that he inflicted on the aid Indians that were apprehended, in order that they should be warned and counsel others not to rebel, but to come and make friends" (353). His account of the execution of the surrendered Tiwas is equally damning—especially given the fact that he himself was a participant: "The soldiers had some of the Indians in a tent" and "were lancing and killing them. He saw that others had been cast into fires, tied, and that some were burned, stabbed and others dead." Cárdenas denied giving the orders but disputes none of the events described, insisting that he

> had been ordered to wage war by blood and fire if the Indians refused to submit, after summoning them to peace . . . because the incident occurred during the excitement and heat of the battle and it was done by all the soldiers. If they exceeded themselves in severity in this case, it was to warn other pueblos in the neighborhood, in order that some would see it and others would hear about it, and none should offer resistance to the Spaniards. . . . It was done thus to spare the other pueblos greater harm, as was the case . . . that the expedition might be carried out more peacefully.

In closing, he stated that he felt sure "that he did not go to excess in anything, nor was there any other principal cause why the Indians revolted, other than their base intentions, and he says that this fact has always been known about them" (364). Later, Cárdenas explained the disappointment of the party:

> The land which this witness has traversed, neither God nor his majesty would have been served by the people of its provinces, as they are disorganized and have no head, and the land is sterile and has no products. . . . On account of the base intentions he discovered in the Indians, and because they have no leader to follow, it would profit little to pacify them and convert them to our holy Catholic Faith. (365)

Unfazed by his total failure, Coronado turned south and found that his abandoned province of Nueva Galicia had erupted in full rebellion. The Mixton War (1540–1541) was a pan-Indian rebellion whose causes and leadership would draw on the same experiences of subjugation, enslavement, and violence at the hands of the Spanish. Never fully settled, the region erupted again in 1550 as Spain fought one of its most costly, lengthy, and militarily unsuccessful campaigns in the Americas: the Chichimeca War (1550–1600). This was yet another pan-Indian rebellion in which Indigenous leaders called for the rejection of Spanish religion and material culture and called for a return to traditional practices. This war was to last until the end of the sixteenth century when Viceroy Villamanrique offered food, clothing, and guarantees of land in exchange for peace (Powell 1952). The region was never militarily conquered.

The Chamuscado-Rodríguez Entrada: 1581–1582

For the next forty years, the Pueblo region remained unattended by Spanish officials. With the ascension of Phillip II to the Spanish throne came important changes in Spain's Indian policies. The New Laws (1542) were written in direct response to and in a direct effort to curtail and control slave raids, forcible labor, and colonial violence against Indians in the mining districts. Unable to enforce these proclamations and facing outright rebellion in Peru and Mexico, the Council of the Indies attempted to reform the practices and legal status of colonizing enterprises by softening and repealing thirty of its provisions against the abuse of Indians. Far from the home country and fearful of losing control over the colony, the Crown was unable to enforce the revocation of the *encomienda* (the practice of exacting tribute) and *repartamiento* (forced labor) systems. In 1573, yet another wave of legislation was issued to colonists outlawing encomiendas, curtailing tribute obligations of Native peoples, and forbidding the illegal enslavement of Indians throughout the colonies (Miller 1985: 111, 117). Even with these proclamations, the Crown sent the colonists mixed signals; viceroys and governors were required to investigate and report to the council and the king the activities and practices of new "discoveries" (use of the term *conquista* had been outlawed).[15] Despite these changes, slave raiding continued to sustain the Spanish need for mining, cash crop, tribute, and infrastructure labor. Enforcement of all of this legislation was nearly impossible in the remote mining regions of the Northern Frontier (Weber 1992: 127–129; Gutiérrez 1991: 149–153; Deeds 1998: 1–29).

Apparently, the organizers of the Chamuscado expedition were unaware of the disappointing reports of Coronado's entrada, but Francisco Gallegos, the chronicler of the expedition, mentions having read a published account of Cabeza de Vaca's epic journey (Hammond 1966: 77). Gallegos was a nine-year veteran of the Chichimeca War, was employed in the service of Governor Ibarra, and had served "partly in the mines of Mazapil, where he was engaged in punishing the hostile Indians who had rebelled, and partly in exploring for mines." Gallegos apparently had also heard from a captured slave about the Pueblos to the north (Hammond 1966: 133).

In June 1582, accompanied by two Franciscan missionaries, eight soldiers, and nineteen Native guides, Fray Agustín Rodríguez and Francisco Chamuscado moved north from Santa Bárbara through the Nueva Vizcaya mining district. Having fought in the Chichimeca Rebellion, Gallegos's account reveals how social relations between the colonists and the seminomadic Conchos, Cabris, Julimes, and Passaguates had worsened. As the result of the actions of recent Spanish slave-raiding parties, the Natives of each of the settlements they encountered had evacuated their villages. After hunting and capturing a group of women and children, Chamuscado reports his attempts to reassure them of their "peaceful" intentions. But he also makes explicit the tactics of violent coercion.

> We also told them not to fear the Spaniards, who would not cause them any further harm, and that we were there for this express purpose—to insure that in the future no Spaniards would come except as their friends, provided the Indians behaved well, but if they misbehaved, the Spaniards would kill them all.

If they wanted to avenge the seizure of their friends, relatives, wives and children, they should come out into the open quickly; for we eight, who were there among them, would avenge the other Spaniards for what they had suffered. Our boldness towards the Natives was meant primarily to intimidate them, so that the news should spread. (71)[16]

At the same time, Gallegos communicates the degree to which the Spaniards perceived of their actions as divinely sanctioned acts. The militarized Christian ideology exported to the Americas helped buttress the outrageous behavior of the colonists in his retrospective account of the actions of the expedition. "It was God's will that instilled fear in these and other Natives, because we know very well that we were not sufficient in number to withstand so many people without the aid of the Lord. With this confidence in divine aid we had set out on the expedition" (72).

Moving gradually northward, the party reached the Piro region in late August, naming the first pueblo they found San Felipe (Hammond and Rey 1966: 81). Finding it recently abandoned, they moved up the Rio Grande and encountered a larger complex of twenty fully provisioned pueblos. Naming the closest pueblo San Miguel, they "found no inhabitants. They had left the night before because they had noticed our approach" (82). Tracing the route of the river northward through the Piro province, they were cautiously approached at an unnamed pueblo "where the inhabitants received us by making the sign of the cross with their hands as a token of peace." The party was given provisions. From here, Chamuscado inquired about the presence of silver at a pueblo he named Malpartida (possibly San Marcos) and was shown a nearby lead deposit.[17]

Hernando Barrando's account of the expedition states that from here, the party moved up the Rio Grande and counted sixty-one pueblos containing "more than one hundred thirty thousand persons . . . [and to the north were] five pueblos in the said province containing fifty thousand souls . . . and eleven mining areas with extremely rich veins, all containing silver deposits" (142–143). He makes no mention of how he arrived at this figure, but it is reasonable to assume that given Castaneda's unvarnished estimate of twenty thousand in 1540, Barrando's estimate was wildly inflated in an effort to convince his superiors (and readers) that the region could support a colony. Refused escort from the Malpartidans, Chamuscado moved northeast to the Pecos region in search of buffalo.

At a pueblo referred to as Piedra Aita (most likely the Keres or Tano pueblo of San Cristóbal),[18] they demanded food and provisions but were refused: "When our men found that the Indians rebelled at our request for food, our leader, rising from his sick bed, and seven companions armed themselves and went to the pueblo in readiness for war. . . . We wanted them to understand that they had to give it, either willingly or by force." After making similar demands for provisions at other pueblos, Gallegos noted a mounting resentment among the Tewas against their presence:

> The friendship they feigned was due more to fear than to anything else. We remained on our guard lest, being Indians, they should treacherously plan some trick and attack us unawares. . . . It was presumably because of that attitude and the fact that we had asked for and taken provisions from them that they attempted to unite the province in order to seize us by force and kill us. (94)

Despite these tensions, Gallegos renders a favorable economic portrait of the Pueblos and emphasizes in his report the ease by which tribute could be collected from the Pueblos. "All the pueblos thus gave us supplies as tribute; and as they are now accustomed to it, they will not resent giving such tribute when someone goes to start settlements" (94).

Wary of the worsening relations between the colonists and Pueblos and disappointed by the resistance of the Pueblos to their missionary efforts, one of the friars, Juan de Santa María, decided to return without escort to Nueva Vizcaya. Gallegos's confusing account of the departure reveals some of the tensions that existed between the missionary objectives of the friars and the economic interests of Chamuscado's party. Later he articulates the notion that the Pueblos regarded the Spaniards as immortal beings, a mythology that first appears in the writings of Columbus and Cortés.

> When the natives saw that the friar was leaving, they became alarmed, believing he was going to bring more Christians to put them out of their homes. . . . They followed the friar and killed him after two or three days of travel . . . even though the natives told us they had slain the father in the sierra, which we named Sierra Morena, we pretended not to understand. . . . Seeing that we paid no attention to the death of the friar . . . they thought that they would kill us just as readily. From then on they knew we were mortal; up to that time, they had thought us immortal. (Hammond 1966: 96)

Chamuscado's account raises important questions: Why had Fray Juan decided to abandon his mission, and why was he not escorted? Why would the Natives have confessed to the slaying, and if they had, why were they not subjected to the same kinds of punitive measures leveled at the Tewas?[19]

After returning to Malpartida, Gallegos claims that three horses were stolen by people from Malagón (presumably this was the southern Tewa settlement of San Lázaro). Chumascado ordered his men to "bring before him the culprits, either peaceably or by force; and to make some arrests at the Pueblo in order to intimidate the natives" (96). The party entered Malagón demanding the culprits; misunderstanding his requests, the Tewas showered him with food from the roofs of their pueblos. Gallegos refused the peace offering, attacked the pueblo, and staged a mock execution of the suspected Pueblos:

> We warned them to deliver those men to us, because we wanted to kill them or take them to our leader so that he might have them put to death; and we added that if the natives did not give up the culprits, we would have to kill them all. . . . Then we soldiers attacked the pueblo again in order to capture some Indians. . . . After apprehending them, the soldiers took them to the camp of the leader to be punished, in view of their crime and as an example to others. Before this happened and before returning to camp, we decided to set fire to the pueblo so that the inhabitants would not perpetrate such a crime again. (Hammond 1966: 98)

After staging a mock execution in which the friars "rescued" the Pueblo men, Gallegos received food and reassurances of friendship. Gallegos's narrative reinforces the tactical use of violence as a means of establishing the subordination of the Pueblos. "Because of what we had done and proposed to do, the natives became so terrified of us that it was surprising how they trembled. This was willed by God on high" (98). Eventually, Spaniards were forced out of Malpartida. Determining that they were no longer welcome in the region and fearing a larger rebellion, they decided to stage a preemptive assault.

> A few days later the Indians assembled for the purpose of killing us, but that did not deter us from going to explore the land in order to verify the information that we had been given. When we left, and again after we returned to camp, we realized clearly and definitely that they wanted to kill us, and that the people of the entire region were gathering for that purpose; so we decided to take precautions and to continue keeping careful watch. . . . In spite of their fear, we came to the conclusion from their conduct that that they wanted to kill us; wherefore we determined to attack and kill them, and to burn some of their small pueblos even though we should perish in the attempt, in order that they might fear the Spaniards. (Hammond 1966: 99)

Gallegos is vague as to whether he did actually carry out the plan or why the friars wanted to remain in light of the reported hostility. There are other questions raised by his account: had they actually burned pueblos and carried out executions, or had they merely made threats? During their stay, Gallegos's men apparently visited or were told about the Acoma Zuni, but he mentions nothing of the character of these encounters.

According to Gallegos, before abandoning the pueblos, the two remaining friars decided to remain in a Tiwa village referred to as Puaray.[20] Chamuscado died of an unknown illness on the journey home. Upon the return of the party, Gallegos was detained by authorities in Santa Barbara over the legality of his entrada. His *relación*, serving as the basis for his defense, was later presented to the viceroy. Accompanying his report is a list of pueblos in the area, although Gallegos makes no mention of having visited any of these settlements in his narration. Like all entrada population estimates, there is no information regarding how the survey of the pueblos was made, the nature of any interactions, or the events surrounding these encounters (how he was greeted, how he communicated with the Pueblos, or whether there was resistance of any kind).[21] Based on Gallegos's report, Viceroy Lorenzo Suarez de Mendoza requested an official commission for exploration and settlement in the area (Hammond 1966: 123). Meanwhile, miners and friars in Santa Barbara made preparations for an unsanctioned expedition to the region on the pretext of trying to secure the safety of the remaining friars.

The Espejo Entrada: 1582–1583

Upon returning to Santa Barbara, Gallegos's accounts of the Pueblos quickly circulated. Hearing these reports, two cattle ranchers, Antonio de Espejo and his brother, Pedro Muñoz de Espejo, offered to lead an expedition north to assist and "rescue" the Franciscan priests. As escaped fugitives from Mexico City, both had been convicted of murdering ranch hands in two separate incidents. Both had fled to the mining districts following their trial. The two brothers curried the favor of Franciscan officials, and in November 1582, the party of fourteen soldiers and two Franciscan priests left Santa Barbara for the Pueblos.[22] Recorded by Diego Pérez de Luxán and Antonio de Espejo, the events of the entrada further demonstrate how violence was routinized and deployed as a means of establishing superior and subordinate relations between the Spaniards and the Pueblos. As is the case with each entrada, the colonists' presence led to abandonments in anticipation of or in reaction to the use of violence as an instrument of intimidation.

Some of the accounts are useful in explaining how the Pueblos were alternately depicted as economic resources and willing parties to the subordination of the colonists. Equally revealing are the references to Puebloan religion as "devil worship." In imagining and representing the Pueblos as religious deviants, the Spaniards simultaneously justified the use of force and affirmed their own moral and religious dominance.

By this, the third entrada, the Pueblos likewise present routinized behaviors designed to shield their communities from the activities and predations of the colonists. Anticipating their behavior, many fully provisioned Native communities were deserted prior to contact with the colonists. At other pueblos, women and children were taken to unnamed and unknown mountain villages, and resistance was offered within heavily fortified pueblos; other villages tentatively offered tribute and food, but they invited no permanent residence. Communication networks appear to have warned the Pueblos of the advances and movements of the soldiers. These behaviors support the notion that in the minds of the Pueblos, the Spaniards as a group had come to represent a single, ethnic category whose behavior could be predicted with great accuracy. In this sense, ascription is an ongoing social process that assigns meaning and signification to groups in a dialectical process.

Upon leaving the mining region, Espejo's party entered territories inhabited by several semisedentary farmers. Referred to as Passaguates, Cabris/Julimes, and Otomoacos, these groups had been subject to some of the most intensive slave raiding in the region. Espejo reports being attacked near the home settlement of one of Espejo's slaves and encountering several recently abandoned villages near the confluence of the Rio Grande and Conchos rivers:

> The Otomoacos had sent word to them that we were going to seize them, and on that account they had taken to the hills. . . . There were a few old Indians in the ranchería, as the others were in the sierra having fled in fear. . . . We decided to go from there to pacify those pueblos, especially some that were known by the interpreter and Gaspar de Luxán, who the year before had been there by commission of Juan de la Parra, Captain of Indehe, to take captives. (Hammond 1966: 162)

Traveling through an area south of contemporary El Paso, they report being well received by several communities. Near Elephant Butte, they noticed "many smoke columns" in the mountains and came across an abandoned village "in ruins" identified by Hammond as San Felipe (Hammond 1966: 171). Moving north along the river (presumably among the Piro Pueblos), they found four recently abandoned pueblos; at one of these (possibly San Miguel), they were greeted by a small party and given provisions. On February 10, they reached a separate province (probably Tompiro), where they observed the village being evacuated with "some small pueblos and many deserted ones" (176).[23] Searching for the Tiwa Pueblos, they were informed by the Tompiros that the province was "in revolt and secretly armed" (174). Finding two large pueblos deserted (yet similarly fully provisioned), they named these Los Despoblados (most likely Isleta), where "the people had fled to the Sierra through fear, thinking they would be killed for having murdered the friars" (176).[24]

Gathering provisions from the pueblo, they found two more deserted pueblos (Los Guajolotes), as well as thirteen other abandoned pueblos, one of which they identified as Puala. Upon learning that the Tiwas had fled into the mountains (Sandias) in midwinter, Espejo attempted to return the Pueblos to their homes:

> When the natives saw us, they fled into the sierra, where we saw seven or eight thousand Indians. We appealed to them in a friendly way. . . . Then some came down and asked for peace by means of signs, agreeing to return to their pueblos because they said their women and children were suffering greatly from the cold. (Hammond 1966: 177)

During the nine months that Espejo's party spent in the region (November 20–July 30), the Tiwa Pueblos remained evacuated. Inspecting the rooms of the abandoned villages, Espejo comments, "In every one of these pueblos, there is a house to which food is brought to the devil. . . . These people set up, midway between the pueblos, their artificial hillocks built of stones like wayside shrines, where they place painted sticks and feathers, saying that the devil will stop there and rest and talk to them" (220). Moving farther north, Espejo visited several Keres pueblos (including Cochiti and San Felipe) and mentions several "Hemes" villages. Moving west to Zia and Acoma, they were provided food and clothing. At Acoma, Espejo mentions the presence of Querechos (Navajo-Apaches) who are "like the Chichimecos" (182) attending irrigated fields.

Peacefully received at Zuni, Espejo "found Mexican Indians [and] also a number from Guadalajara, some of those that Coronado had brought" (184), and he assessed the resources at Zuni: "if there are good mines, this will be the best land ever discovered because the people of these provinces are industrious and peaceful" (184). Through translators, Luxán was informed of Coronado's attacks against the

Tiwas and provides evidence supporting the enslavement of the remaining Tiwas by Coronado's party:

> Coronado . . . pressed them so hard that those people who did not die at the hands of the Spaniards—Coronado's people, whom the Indians called Castillos—died of hunger and thirst. . . . Finally, the people of Puala surrendered and threw themselves on Coronado's mercy, and he took as many, both men and women, in his service as was necessary, and returned to this pueblo. From here he set out for the valley of Samora, which must be one hundred leagues distant from this province. (Hammond 1966: 184)

In the only reference to any possible epidemics, Luxán states that there was no evidence for diseases. "The people here are extremely healthy, for neither in this province nor the others we have crossed have we seen any sick or crippled persons, but only many old ones" (186).

Indeed, Luxán was intent on creating a "new" Mexico in the minds of the officials. For him, the Indians here were "better" than Mexicans: "men and women are pure Mexican in the way they walk, cry and even in their dwellings, but neater than the Mexicans" (186). Estimating the Zuni population at twenty thousand, Espejo journeyed east. Intent on finding silver deposits, they organized a party to visit the western lands of the Hopi. Informed that the Hopi had similarly fled the region, Espejo requested an escort from Zuni. "There was a great gathering of wild and warlike people to oppose us; that the other pueblos were awaiting developments since they had not decided either for war or for peace; and that the children, women and girls were in the sierra with their flocks" (Hammond 1966: 187).

Negotiating passage through their Zuni interpreters, Espejo's party reached the Mohoce (Hopi) settlements without incident. In describing the religious practices of the Pueblos, Luxán states:

> They worshipped the Devil. Here in this pueblo of Aguico [Háwikuh] and in the others are some small prayer houses where the Indians speak to the devil and give him offerings of ollas and earthen bowls containing pinole and other vegetables. Father Fray Bernardino erected a cross next to one of them, and when some of the servants went inside to get earthen bowls, the old men said the devil was no longer in the house, that he was angry because of the cross which the Christians had erected, and because they had entered his house. (186)

Espejo's glowing reports of wealth and tribute-rich populations are contradicted by Luxán. In his report to the viceroy, Espejo claims to have been told of vast ore deposits and wealth in regions farther north. "A Cacique and some other people told us that they knew about the above mentioned lake where there were riches of gold" (Hammond 1966: 227). While Luxán characterizes the mining deposits of the region as "so worthless that we did not find in any of them a trace of silver, as they were copper mines, and very poor" (197), Espejo glowingly reports, "I found the mines and took from them with my own hands ores which, according to experts on the matter, are very rich and contain a great deal of silver" (227).

In contrast to Espejo's evaluations of the region's resources, Luxán communicates a different motive for the decision of the friars to abandon the expedition and return south to Santa Barbara; he also articulates the need for profitable economic ventures to sustain the colony:

> We decided . . . since we had not found mines aside from one lone discovery—that we should go together to the provinces of the Tiguas and Maguas and look for them, since these provinces were situated on our route and there were reports of mines there. For it was evident that if there were none, the land could not be settled nor all of these many souls saved. (Hammond 1966: 198)

Espejo's population figures at Hopi (fifty thousand) are similarly contradicted by Luxán, who suggests a much smaller figure (twelve thousand).[25] These disparities are quite significant and are a consistent feature of each entrada narrative. Clearly, those who supported colonization habitually inflated population figures in an attempt to impress their readers and garner support for a colony. The fact that many of these figures are used by scholars to support large-scale population crashes (motivated by diseases that are never mentioned) is difficult to understand, given this contextual meaning. Deciding to return to the Tiwa region, Espejo characterizes the departure in glowing terms: "When we took leave of them, they made many promises, asking us to come back again and to bring with us a large number of 'Castillos,' their name for the Spaniards; for in this expectation they were planting a large corn crop that year in order to have enough for everybody" (228).

Moving south to Acoma, Espejo's party was involved in an altercation with the Querechero village at the base of the mesa at Acoma and "found the people of the pueblo in rebellion" (Hammond 1966: 200). Determined to reenter the Tiwa region and find silver deposits, Espejo returned to the Tiwa region in late June (203). At Puala, the soldiers orchestrated yet another punitive attack against the Tiwas:

> Efforts were made to have the Indians come peacefully to us, but though they said they were our friends, they would not bring their women to the pueblos. Instead they remained in the sierras and mocked us. Seeing that unless we

administered some punishment, they would soon become impudent and try to kill us, we determined to do what I shall, record below. . . . When we asked them for food, as they were our friends, they mocked us like the others. In view of this, the corners of the pueblos were taken by four men, and four others with two servants began to seize those natives who showed themselves. We put them in an estufa [kiva]. As the pueblo was large and the majority had hidden themselves there, we set fire to the big pueblo of Puala, where we thought some had burned to death because of the cries they uttered. We at once took out the prisoners, two at a time and lined them up against some cottonwoods close to the Pueblo of Puala, where they were garroted and shot many times until they were dead. Sixteen were executed, not counting those who burned to death. . . . This was a remarkable deed for so few people in the midst of so many enemies. (204)[26]

Unable to induce the Tiwas to return to their pueblos, Espejo journeyed east to Galisteo and San Cristobal. Finding the pueblos fortified and refused provisions, the expedition turned to the south in early July, returned to the mining territories of Santa Barbara in late September 1583 (208, 212).

Espejo was ultimately arrested and charged with fleeing Mexico City after his arrest. Despite these concerns, Espejo requested an official commission from King Phillip II for the pacification and settlement of New Mexico (Hammond 1966: 233). Espejo's letter belies the discrepancies between Luxán's report concerning the resources and dispositions of the Pueblos following his entrada, and it demonstrates once again the contexts in which hugely inflated population estimates were generated:

What I am reporting represents only the smallest part of what exists in those provinces; for as we traveled through them we received reports and information about large towns, very fertile lands, silver mines, rumors of gold, and peoples better governed, all in the direction of our march. . . . The natives of all those provinces are large, more vigorous than the Mexicans, and healthy, for no illness was noted among them. . . . There are many rich mines, too, from which I brought ores to have assayed and to determine their quality. (230)

He also mentions having captured a "Tamos" man and a "Mohoce" (Hopi) woman from the region, "so that they might enlighten us regarding those provinces and the route to that region. . . . My intention being that these two Indians should learn the Mexican language and other tongues with this aim in view" (Hammond 1966: 231).

Competing with him were Hernán Gallegos and the then lieutenant governor from Nuevo Leon, Gaspar Castaño de Sosa. Eventually, the license was awarded to Don Juan de Oñate, a wealthy mine owner from Zacatecas. While awaiting official permission, Castaño de Sosa staged an illegal expedition to pacify (for the fourth time) the Pueblos of New Mexico.

Castaño de Sosa-Morlete Entradas: 1590–1591

According to Hammond, disappointed with diminishing returns of the mines in the area and hearing the favorable accounts of Espejo's expedition, Castaño de Sosa abandoned his province, uprooted his entire colony, and set out for New Mexico in late July 1590 (Hammond 1966: 29). Warned by the viceroy in June of the same year to discontinue "taking and selling Indians as slaves . . . for use on the farms and in the mines of Northern Mexico," Castaño de Sosa departed, "leaving behind many natives who were deeply grieved by his departure because of their association with the Spaniards and their association with them" (from Castaño de Sosa's *memoria*, memorial or memoir, cited in Hammond and Rey 1966: 245).

Accompanied by 170 members of the colony (he does not mention the number of Indians, but livestock included horses, sheep, and cattle), Castaño followed a meandering path north, choosing to explore the region from the eastern portion of the Rio Conchos and Rio Grande (Hammond 1966: 253). Their first encounter was at a place called Río de los Nadadores. Here, two Indians were accused of stealing horses. Castaño's reprisals reveal the general disposition of the colonists toward the protection of their property as well as the use of aggression and brutality as a means of maintaining dominant social positions among previously enslaved Natives: "As a warning to the rest of the Indians in the camp, as well as to admonish the natives of the region through which the army would later pass, Castaño de Sosa ordered that two of the thieves be hanged; the other, who was very young he entrusted to a soldier in the company" (Hammond 1966: 246).

Castaño's narration is difficult to follow; as a result, we only know that from the junction of the Conchos, he followed the Rio Grande north on the east side of the river. Believing that there were larger settlements on the outskirts of the Pueblo region, he attempted to enter the Pueblo region from the east by following the Salado (Pecos) as it flowed into the Upper Rio Grande valley. The path was treacherous and progress was slow.

Four months later (on November 20), Castaño notes observing a column of smoke in the hills near the region of the Guadalupe

Mountains (near Carlsbad), and on the 29th, his men mention finding a settlement on the banks of the Pecos, "that it was inhabited and that he thought the people were leaving it" (262). On December 2, Castaño sent an advanced party to "find some Indians and bring them back to camp . . . but not to enter any town if they found one" (263). On the 23rd, the advanced party returned from Pecos and informed Castaño that they had found the pueblo and had entered it to escape from the snow and acquire some food.

According to Castaño, they were well received and fed at first, but they were attacked when they attempted to reenter the pueblo and collect more food the following day, "leaving in the hands of the Indians five harquebuses, eleven swords, nineteen saddles, nine sets of armor, and a quantity of wearing apparel and bedding" (Hammond 1966: 267). Castaño returned to Pecos on the December 30:

> When he reached the town, he noticed that the natives were in battle array, men as well as women standing fully prepared on the terraces and down below. . . . He called to the Indians and told them he would not do them any harm or injury, but this failed to calm them. They attempted to gain entry for five hours; the captain of the pueblo was given a knife and some other trifles, but even this failed to pacify the natives. . . . The men repeated that the natives should be overcome by force since they had refused to accept the friendship we had offered in peace and goodwill. (270, 272)

After an extended battle, Castaño met the leader of the pueblo and attempted to communicate his demands; "by means of signs, the governor told the chieftain that . . . he had not come to harm them in the least, and that they should not be afraid. The natives understood him clearly and they soon brought out some food which they threw down on us from the corridors" (275). After spending the night near the pueblo, the colonists discovered that the pueblo had been evacuated during the night: "At dawn the next day, not a single inhabitant was to be found in the town, a development which distressed us all greatly when they heard of it. . . . They had abandoned their homes in the bitter cold of winter, with its strong winds and heavy snows, conduct which seemed incredible to us. Even the rivers were frozen" (279).

After his soldiers located and captured a small group of fleeing Pueblos, Castaño interrogated his captives. "They had indicated by signs that there were such ores in the other pueblos mentioned earlier; and on hearing this we decided to look for the mines" (Hammond 1966: 279). Castaño claims to have visited several unnamed pueblos along the way (most likely the Tewa villages of Tesuque, Nambé, Cuyamungué,

Pojoaque, and Jacona and San Juan). In each pueblo, a similar scene was repeated, signaling to the Pueblos the ceremonies of possession required by Spanish law. More significantly, whether the Pueblos understood the message or not, Castaño established a legal claim on the Pueblos and justified the use of force (and enslavement) of those who resisted. This practice, common on the Northern Frontier, facilitated the alienation of rights afforded to Indians by the New Laws. Castaño records having performed the ceremonies of possession at each of the pueblos he encountered—setting the stage for a legal defense should his activities be investigated:

> The next day, the lieutenant governor ordered us to erect a tall cross, which we did, raising it to the sounds of trumpets and harquebus shots. The Indians swore obedience in the name of our king and sovereign and the lieutenant governor, accepting their allegiance, appointed a governor, an alcalde and an alguacil, all in the name of his Majesty. (280)

Moving south and west through the Tewa Pueblos, the colonists observed the evacuation of a pueblo:

> When we drew near this pueblo we saw many people leaving it. . . . Our leader explained by signs that the Indians should not run away, because he had not come to harm them, but on the contrary to protect them in the name of his Majesty and to be their friend. All this was indicated by sign language, but they understood him clearly and had no objections to offer. . . . They pledged obedience to his Majesty. (282)

After viewing eight more Tewa pueblos, Castaño arrived at a deserted village similar in description to San Juan, where

> the very sight of us frightened the inhabitants, especially the women who wept a great deal. In view of this we circled the pueblo once, but no one appeared. . . . All agreed to camp for the night at some huts . . . where there were people from other places who had come to trade with this settlement. . . . As we approached the settlement, the Indians, both men and women, began to move away. . . . It seemed as if the lieutenant governor was especially endowed by God to win allegiance to his Majesty from these barbarians, so that by the divine will they might be brought to a knowledge of the Catholic faith. The governor had a cross with an image of God our Lord, and whenever he came to a pueblo he held it in his hands while he and his companions knelt reverently to kiss the crucifix, at which the barbarians marveled. (Hammond 1966: 286)

From here, Castaño's party moved south, reading the terms of possession at seven "Quereses" pueblos (the Keres pueblos of Cochiti,

Santo Domingo, San Felipe, Santa Ana, and Zía on the Jémes River). Moving southeast through the Galisteo Basin and camping at the Tewa pueblos of San Cristóbal, San Lucas, San Marcos, and San Lázaro, Castaño revisited Pecos without incident (288–289). From Pecos, he resumed his search for mines in the still abandoned Tiwa province, sending the remainder of his party north to Santo Domingo. While exploring the region, Castaño's chronicler notes the presence of twenty (of the twenty-one) recently abandoned pueblos, where "the Indians said that most of them had been abandoned by the inhabitants due to fear and that they had sought refuge in the mountains or in other pueblos" (293). Castaño narrates:

> These were the pueblos whose people had killed the friars reported to have come this way. . . . While we were there, we noticed that the one across the river was being evacuated in part, and in order to prevent its complete abandonment . . . we . . . forced some of the people who were fleeing to turn back. . . . Our leader reassured the natives, giving them to understand that they should not desert their homes. (293)

Moving back north to Santo Domingo, Castaño located six additional recently abandoned pueblos, one of which was partially inhabited. Returning to Santo Domingo in late March, he was arrested by the official representative of Viceroy Luis de Velasco, Juan Morlete, for launching an illegal expedition and participating in the illicit slave trade in Nuevo León.

According to the testimonies of other members of the party, Castaño's party collected slaves throughout the entire trip in an effort to placate the members of his party.[27] Upon learning of Castaño's illegal foray, Viceroy Velasco sent a military expedition north to capture and return Castaño de Sosa to Mexico City. According to then viceroy of New Spain, Villamanrique, both Luis de Carvajal's (the deposed governor of Nuevo León whom Castaño had temporarily replaced) and Castaño de Sosas's participation in slave raids were cited as the primary cause of the Native rebellions in Nuevo León. The letters from Viceroy Villamanrique to his successor, Don Luis de Velasco, reveal the scale and nature of the frontier Indian slave trade as well as the attention devoted to the problem at the highest levels of the Spanish colonial government:[28]

> In the kingdom of New León, a certain Luis de Carvajal de la Cueva, formerly served as governor. It was his custom to go inland with a company of evil and unscrupulous men, who would go northward to the Rio Bravo and Rio de

Palmas, where the Indians had never seen any Spaniards nor committed any crimes. Then, like someone hunting rabbits or deer, the soldiers would seize each time from eight hundred to a thousand Indians and sell them in Mexico. Because of this practice, the Indians became angry and began to resist, even to the point of stirring up the natives here to make war. (Hammond 1966: 296)

Later, the viceroy explains how the arrest of Carvajal had been made after hearing for several years of the problems caused by the practice:

The fiscal had informed the Audiencias of this crime before I came. Later, after pursuing this case and many others that came before me, I declared that all the Indians who might be taken should be free, and not subject to slavery, and that generally from here on the captains would not be permitted to sell the Chichimecos Indians as slaves, and I so ordered. . . . My order was the principal means of bringing the Indians into their present state of peace; and because Luis de Carvajal disobeyed it, he was summoned to appear before me. . . . Paying no attention to this prohibition, he returned again to the region in question, where he once more began his criminal seizures of Indians. (297)

Soon after, Viceroy Velasco sent Captain Juan Morlete into New Mexico to retrieve Castaño. His orders included the immediate release of Indian slaves in the frontier:

You will try to gather together all the Indians you encounter, male or female, regardless of age, who have been sold or despoiled of in that region . . . and you will leave them along the route at the spots they recognize as their places of origin . . . explaining to them that your principal object in coming was to free them and restore them to their own lands, and to punish those who had troubled and injured them. (Hammond 1966: 299)

The purpose of Castaño's entrada is further clarified in a Velasco's instructions to Morlete; apparently his expedition was funded by officials in the border regions. Velasco was also apparently unaware of the findings of Coronado's entrada fifty years before and instructed Morlete to gather information regarding the resources of the Pueblos:

In view of the information I have received that certain evilly disposed persons, moved by greed, aided Gaspar Castaño de Sosa and his men with provisions and equipment in exchange for a guarantee of repayment in slaves, . . . you must warn everyone you meet that you will take away from them all the slaves they have seized and will return these slaves, as you proceed to the places whence they were removed or wherever they wish to go. . . . You will note in all the territory through which you travel . . . whether there are farmers among the natives, whether the people are industrious or lazy like the Indians here, and also the nature of the land in regard to the mining prospects. (299)

After receiving his instructions, Morlete moved north with a party of forty soldiers and a priest. After reaching Castaño at San Felipe, Morlete arrested Castaño de Sosa and spent a month exploring the region.[29] Although his account provides little information regarding the pueblos, Morlete reported finding:

> several deserted pueblos. It is said that the inhabitants fled to the mountains when Castaño and his men came to the land, since they feared that the Spanish would harm them in revenge for the death of the friars who were killed in this region. . . . I felt keenly the lack of interpreters who would enable me to talk with the people and reassure them, but I did what I could by means of signs. (303)

Although the objectives of Morlete's expedition were ostensibly limited to the capture of Castaño, Morlete's report indicates the interest of Spanish colonists in extracting labor and tribute from the Pueblos—a sentiment shared by Viceroy Velasco.[30]

Permanent Colonization: The Don Juan de Oñate Entrada of 1598

The final recorded entrada into New Mexico was undertaken by Don Juan de Oñate, the eldest son of Don Cristóbal Oñate, the former governor of Nuevo Galicia. Don Cristóbal, an immigrant from the Basque region of Spain, had served under Cortés in the siege of Tenochtítlan. While acting as Juan Vázquez Coronado's lieutenant during Coronado's entrada, Don Cristóbal had been appointed governor during the Mixtón Rebellion of 1540 and had helped establish some of Spain's most productive mines in the Zacatecas region (Powell 1952: 12). Don Juan de Oñate's membership in the newly wealthy "silver aristocracy" of Zacatecas, as well as his political associations with Viceroy Don Luis de Velasco, enabled Oñate to secure the legal contract for the pacification and settlement of New Mexico in 1595 (Hammond and Rey 1953: 6, 59). Appointing his nephews Don Juan de Zaldívar as *maese de campo* and Vicente de Zaldívar as *sargento mayor*, Oñate's venture was intended to establish self-sustaining and profitable permanent colonial presence in New Mexico.

Believing that New Mexico was a land rich in mineral wealth, tribute, populations and agricultural surpluses, Oñate offered to fund the expedition at his own expense (Hammond and Rey 1953: 70). His letter to King Phillip II reveals the grandiose expectations Oñate and his followers had regarding the province's wealth and tactical significance. Oñate requested full control over the appointment of encomiendas and tribute levels provided by the Pueblos.[31] More significantly, he hoped to circumvent the control of the viceroy in Mexico City. Requesting full viceregal control over the region and the title of marquis, Oñate sought full rights regarding the processing and minting of silver currency (47–57, 489). Unaware of the geography of the region, Oñate asked that he be "allowed to bring two ships per year, free of taxes and import duty, to provision the land and exploit the mines that may be discovered" (34). For three years, Oñate waited for official license to be granted. By 1596, the Council of the Indies began to question Oñate's credentials as well as his financial motives for the expedition.

In a letter from the Council of the Indies to King Phillip II, the council recommended against Oñate's appointment in New Mexico, suggesting that his personal debts would compromise the success of the colony: "[Oñate] has so mismanaged his estate that he now owes more than thirty thousand pesos in bad debts . . . that it will be impossible for him to attain the objectives of this expedition . . . that the only people who will join him will be the outcasts and vagabonds, who serve only to cause disgraceful disturbances and riots . . . this has already been experienced in the entrada made by Captain Castaño" (Hammond and Rey 1953: 90).

Despite these reservations, by January 1598, Oñate had assembled 120 soldiers, 10 Franciscan priests, 43 carts, and nearly 6,000 head of livestock for the journey to New Mexico.[32] Included in his party were a Tewa woman (whom Oñate refers to as "a second *Malinche*") and a man from Pecos who had been captured as slaves during Castaño's entrada.[33]Although he mentions no contact with any Native communities, on April 30, near El Paso Texas, Oñate claimed possession of the lands of New Mexico. Moving up the eastern side of the Rio Grande, he encountered two of the southernmost of the Piro settlements recently abandoned.[34] Camping for a month near a pueblo named Teypana, they were greeted and fed by a Cacique but found the remainder of the pueblos in the region abandoned (318).[35]

Passing through the sparsely populated Piro settlement of Sevilleta and the recently evacuated San Juan Batista, Oñate was visited by a party of Pueblo representatives. Included among them were Indians brought north by earlier expeditions.[36] With the aid of these messengers, Oñate moved gradually north. Finding several uninhabited pueblos in the Tiwa region, "on account of fear," Oñate visited Puaray, Santo Domingo, Zia, and set up a capitol in the Tewa pueblo of San Juan in mid-August (Hammond and Rey 1953: 320).

In the following months, Oñate's party contacted Picuris, Taos, San Ildefonso, San Cristóbal, Santa Ana, Pecos, Zía Acoma, Zuni, the Hopi region, as well as Jemez. No account is given of how they were received at each of these pueblos, although he records having read the Requerimiento and appointed governors at Santo Domingo, San Juan, Acoma, Zuni, and Hopi and appointed seven priests to serve in each of the provinces (Hammond and Rey 1953: 17–18, 329–360). Forcing the Tewas to provide food and shelter in the newly established capitol at San Juan (Yunque-Yunque), the Tewas were resettled in an adjoining village after building a chapel for the Franciscan missionaries (Ellis 1989: 5–23).[37]

Almost immediately, the colonists registered their disappointment in the poverty of resources in the region, and while Oñate was surveying the southern pueblos, several members of his party deserted the colony (323). In a letter to the viceroy, Oñate records, "The mutiny of more than forty-five soldiers and officers, who in anger of not finding bars of silver on the ground right away and resentful because I did not allow them to abuse the natives either in their persons or property, became dissatisfied with the land, or rather with me" (481).

Declaring the mineral wealth and large tributary populations among the Pueblos (which he estimated at sixty thousand), Oñate promised to deliver to the king "a New World, greater than New Spain" (492). The reports of his deserting captains tell a different story. Written from San Juan in March 1601, Captain Luis Gasco de Velasco recounts how the colony was sustained for three years through the forcible acquisition of food and clothing—acts that contributed to a mounting resentment against the colonists:

> Because of this and other annoyances, the Indians fear us so much that, in seeing us approach from afar, they flee to the mountains with their women and children, abandoning their homes, and so we take whatever we wish from them . . . until this year this tribute has been collected with such severity that [they] had nothing but what they had on [their backs]. . . . The Spaniards seize their blankets by force, sometimes even when it is snowing, leaving the poor Indian women stark naked. (609–610).[38]

In the following year, Oñate intensified his efforts to ascertain the mineral wealth of the region. Leaving the capitol for months at a time, Oñate's forces were involved in a succession of conflicts among the

Pueblos. The tactics and methods of coercion and retribution employed by Oñate's party had achieved a measure of tenuous control over the Native communities in the mining districts, but with renewed efforts by the Crown and viceroy to reform the frontier Indian policy, Oñate's behaviors in New Mexico attracted both attention and criticism at the highest levels of Spanish government.[39]

In addition to the "unofficial" foraging of the colonists, tribute demand included "two thousand cotton blankets, one and a half yards square, five hundred buckskins, two thousand fanegas of maize [one *fanega* equals approximately one hundred pounds],[40] some beans and fowl" (638). This figure is difficult to imagine. Based on these measures, each pueblo would have had to provide *twenty thousand* pounds of corn per tribute period—an astounding burden given the natural limitations imposed by drought and severe winters in New Mexico.

In letters written to the viceroy, Oñate's lieutenant governor Peñalosa and Fray Francisco de Zamora record the responses of the Pueblos to these demands. During the first two years, the colonists had exhausted several years of surplus corn collected by the Pueblos as insurance against droughts. One of the priests narrates:

> Your lordship's chaplains, the royal alférez, and all the other captains and soldiers reflected on the fact that we had consumed all the corn that the Indians had saved during the preceding six years, because there has not been a week since we came here that we have not used up from fifty to sixty fanegas of corn, and when the governor and the rest of the people were here, we consumed upwards of eighty fanegas. As a result of this, the Indians have reached the state of famine I described above. (Hammond and Rey 1953: 696)

As the result of the tribute levels, the region was rapidly being depopulated:

> The Indians [have] run away. This witness has seen many pueblos abandoned, the people having fled for fear of ill treatment. . . . The fact is that in order to induce the Indians to furnish corn and food, it has been necessary to torture the chieftains, even to hanging and killing them. . . . We . . . see the natives starving to death, eating whatever filth there is in the fields, even the twigs from the trees, dirt, coal and ashes. This witness knows that many of them have died of hunger . . . not to mention . . . the treatment of the Indian women. I know for certain that the soldiers have violated them often along the roads. (674)[41]

Sending parties to the east, south, and west, Oñate learned while among the Zunis that his nephew and several soldiers had been killed while attempting to acquire provisions from the people of Acoma. In retribution, Oñate ordered that a "total war" of fire and blood be waged against the Acomitas, and the pueblo was assaulted in late January 1599 (Hammond and Rey 1953: 614). Claiming to have been aided by the appearance of Santiago, the colonists burned the entire pueblo. After three days of battle, the Acomitas surrendered (Hammond 1953: 326, 427). Eight to nine hundred Pueblos were reported killed. After the survivors fled to the safety of the kivas, Oñate's sargento mayor Vicente de Zaldívar

> ordered these prisoners brought out one at a time, and as they came out . . . put them to the sword and hurled them down the rock on which the pueblo is situated. . . . Then the sargento mayor ordered the estufas and living quarters to be set on fire. Many were burned alive in these places, men and women, some with children in their arms; others were suffocated by the smoke. (614)

Oñate led the remaining captives to Santo Domingo where eighty men and five hundred women and children were sentenced for the insurrection. Oñate ordered that all males over twenty-five were to have one foot cut off and sentenced to twenty years of slavery; women were sentenced to twenty years of servitude. The remaining boys were remanded to the "custody" of Vicente de Zaldívar, and the girls were assigned to the servitude of Commissary Fray Martínez. For unknown reasons, the elderly were turned over to a unnamed group of Jumanos (475–476). Two similar punitive attacks occurred against three Tompiro villages (one of which may have been Abó) as well as at the Jumano settlement.[42]

Most likely, these and other captives were relocated to serve the colonists in the colonial capitol at San Juan. The location of these events is most significant. Because San Juan served as the home of the colonists, it seems probable that many of the prisoners taken in raids and battles (especially at Acoma) lived in close proximity to the Tewa people at San Juan. Given the fact that Po'pay's rebellion was initiated at San Juan, the physical presence (and memories) of these Pueblos may have helped initiate the beginnings of the pan-Puebloan movement. Recounted through oral histories, the experiences of the various Pueblos captured and held at San Juan may have served as the basis for an emergent discourse of resistance.

Faced with the desertion of some of the most trusted friars and soldiers of the expedition, Oñate returned to Nueva Galicia and appealed to the viceroy for reinforcements. By this time, the viceroy had considered the full letters and testimonies of Oñate's supporters and detractors.

Finding only traces of copper in the assays sent by Oñate, Viceroy Montesclaros wrote a letter to King Phillip indicating his disbelief in Oñate's exaggerated accounts: "I cannot help but inform your majesty that this conquest is becoming a fairy tale" (Hammond and Rey 1953: 1009).[43] By 1606, Oñate was recalled to Mexico City by Phillip II (1036). Forced to resign his commission, Oñate and his remaining soldiers were charged with "excesses and transgressions committed in the entrada, conquest, and government of New Mexico" (1109).[44] New Mexico was to be maintained exclusively as a missionary field colonized by a group of fifty married colonists and six Franciscan missionaries (1073, 1077).

Colonial Period Conflicts: New Mexico 1610–1680

Documentary evidence regarding the activities and relationships between the Pueblos and Spanish colonists in the seventeenth century is largely fragmentary. Following the capture of the capitol at Santa Fe in 1680, the Pueblos burned nearly all of the extant historical records, letters, and manuscripts. Still, primary documents have been collected by Twitchell (1911, 1914) and Hackett (Hackett and Bandelier 1923). In the 1930s and 1940s, historian Frances Scholes published a series of essays examining the struggles between civil and ecclesiastical officials (1932, 1935a, 1935b, 1937, 1942a, 1942b).[45]

Following the relocation of the colonial capitol to Santa Fe in 1610 and the establishment of the *custodia* (the religious capitol) at Santo Domingo in 1616, missionary efforts proceeded at a slow pace (Hodge 1945: 2). Under orders from the viceroy to relocate the Pueblos into large more easily monitored settlements *(congregaciones)* according to the policies of *reducción*, Governor Peralta was given authority to apportion encomiendas and set restricted tribute levels (Hammond and Rey 1953: 1090–1091). After receiving their assignments to various districts, missionaries visited larger pueblos and rented individual rooms from selected pueblos with varying degrees of success (see table 5.2). Little is known about how the pueblos were approached or how successful these attempts were. One thing is certain; both Pueblos and Spaniards were able to assert social boundaries, regulate interaction and insulate themselves from the influences of the other group (see figure 15). The Pueblos did not simply accept Christianity, nor could they have been fully aware of the doctrines revealed in the gospels themselves.

TABLE 5.2. Mission Construction Dates

Name of Settlement	Church Name	Building Dates
Abo	San Gregorio	1630
Acoma	San Esteban	1644
Chilili	La Navidad	1616
Guisewa	San Jose	1626
Hawikuh	La Purisma Concepción	1630
Humanas	San Buenaventura	1659
Isleta	San Agustin	1629
Pecos	Nuestro Sra. de Los Angelos	1625
Quarai	La Concepción	1633
San Cristobal		1626
San Juan		1600 (est.)
Santa Fe	San Miguel	1640
Tajique	San Miguel	1650
Zia	N.S. de la Asunción	1614

Language served to insulate Spanish rituals through written and spoken forms of sacred speech—Castillian Spanish and Latin.

The Pueblos, too, were probably never fully understood. Unlike the Jesuits, who were interested in conveying the meaning of Christianity through the use of Indigenous languages (and transforming the hearts of Indians), the Franciscans used translators and intermediaries to communicate the meanings of the gospels. These figures, known as coyotes, could move back and forth between the sacred realms of both cultures, and during the Revolt, they were singled out as being especially powerful and influential leaders.

The Pueblos similarly continued to insulate themselves from the prying eyes of the Spanish. At Cochiti, oral histories record the Pueblos moving their dances and ceremonies into the hills, away from the eyes of the missionaries or their acolytes. Trade was similarly based on entirely different assumptions. For the Spanish, tribute was the expected manner of exchange. For the Pueblos, goods were given with the expectation of reciprocity. These two very different systems and the Spaniards insistence on exploitative relationships would lead the colony to disaster during the 1600s.

We simply are not sure of how intensively the friars interacted with the Pueblos at all, or what the terms and conditions of their residencies were, but they seem to have been negotiated locally. At Pecos, the friars

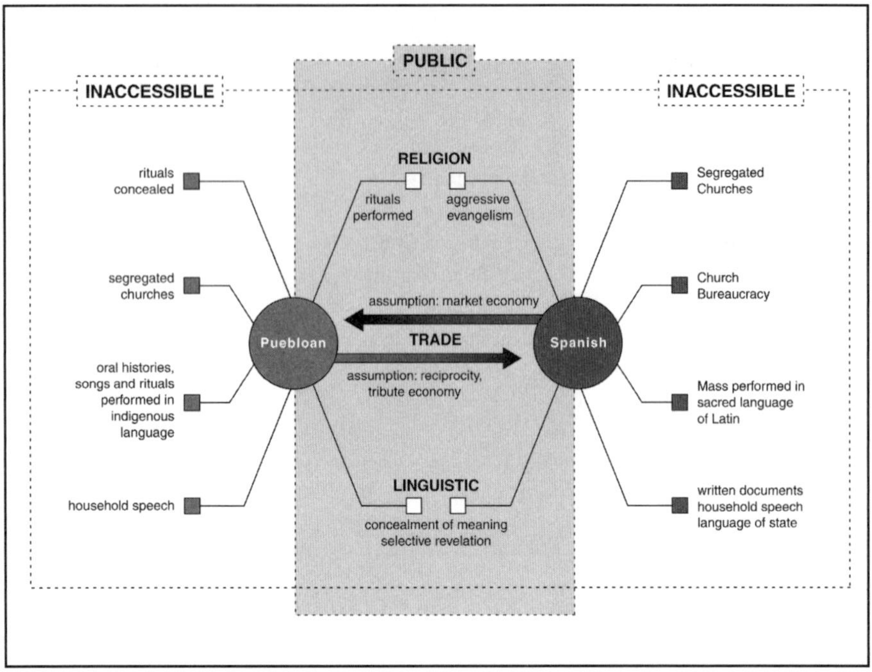

FIGURE 15. Boundaries between Pueblos and Spanish. Traditionally it has been assumed that Indigenous peoples are quickly acculturated or transformed by interaction with colonists. In fact, both sides insulate themselves through an active and conscious process of boundary maintenance. Both groups segregate religious performances by restricting access to sacred languages (literacy was not shared; neither was Latin). Masses were said in separate churches, and the Pueblos concealed ritual behind closed doors. Public performances (feast and katchina dances and Spanish performances of the Reconquista) had a specific intent, but they were likely unintelligible by members of the opposite group.

were not allowed to live in the main pueblo; and given the small number of missionaries, *conventos* (missionary residences) were most likely the places where the friars were allowed to live while in residence.[46] Use of small plots of land by the friars was negotiated, and European methods of agriculture, domesticated animals, and cultigens were gradually introduced. Conventos were built at many of the larger pueblos. Church construction proceeded gradually at a number of pueblos (see table 5.2, after Kubler 1940: 118–127), and by 1628, nearly fifty churches had been built.

At the same time, the colonial population in Santa Fe (served by 700 Indian and mestizo slaves) had swelled to 250. Slave raids had continued, prompting Custodia Parea to lament that Indian children were being snatched from their parents, "as if they were yearling calfs or colts . . . and placed in permanent slavery" (Hackett 1923: 1300). Encomiendas required tribute levels of one fanega of maize (about one hundred pounds), twice per year from each Pueblo household as well as 5.6 square feet of cotton textiles (Gutiérrez 1991: 104–105). Disputes between secular and religious officials over Indian labor were a recurrent theme in seventeenth-century Pueblo-colonial relations. The strategies of survival that had been developed by the Pueblos in order to survive years of drought were severely challenged by the labor and tribute systems introduced in the seventeenth century.

Not only were mining, ranching, textile, and farming communities expected to support the colonists, but the Crown controlled trade in most commodities imported to the New World. Gunpowder, arms, chemical agents necessary for silver smelting, oils, clothing, and agricultural implements were sold at exorbitant prices, placing heavy demands on the financial resources of colonists. Encomiendas established in the early days of the colony competed with the needs of civil and ecclesiastical authorities for the use of Puebloan labor. Missionaries appropriated Indian labor for the construction of churches, conventos, fields, ranches, and gardens. Encomenderos and colonists could demand at their discretion Indian labor for a variety of time-consuming ventures; new fields were cleared, and homes, ranches, and irrigation canals all required constant attention. As the cycles of planting and harvest coincided with the agricultural cycles of the Pueblos, the immediate needs of the Pueblos were largely ignored by colonial officials. Reduced to smaller agricultural fields by *congregaciones*, surpluses intended to sustain the Pueblos during times of drought were collected as tribute.

Individual governors increased the strain on the Pueblos as well. In paying the government fees for the office of governor, heavy debts were incurred. Provided by the Crown, the wagons used to deliver missionary supplies to the colony were returned to Nueva Vizcaya loaded with several tons of textiles, produce, salt, and groups of Indian slaves for sale in the inflated markets of Zacatecas (Scholes 1930: 188). Salt deposits (used in the processing of silver) were heavily mined in New Mexico. The scale of the salt trade is staggering: of the sixteen wagons sent south, up to ten were loaded with salt; at two tons per wagon, each of the biennial supply trains could carry twenty tons of the highly valued mineral.

Missionary Efforts: Forcible Conversion and Native Resistance

Expanding the missionary presence, small groups of Franciscan priests and brothers arrived between 1618 and 1626, bringing the total number of missionaries to twenty-six (Kubler 1940: 7; Scholes 1937: 20). Setbacks were frequent. Zarate Salmeron, whose memoria was written in 1621 as a plea to the king not to abandon the colony, claimed to have baptized 6,500 Jemez Indians by 1625. Salmeron claims that by this time, 34,650 Indians had been baptized. Like all baptismal figures, these numbers were likely exaggerated in an effort to convince the king that there were simply too many converted Pueblos to abandon the region and that if he did, the king would be risking his mortal soul. Jemez, the supposed site of Salmeron's greatest success, was burned in 1632 and abandoned completely by 1639 (figure 16). His replacement, Fray

FIGURE 16. The Jemez Mission of Guisewa. The Mission of Guisewa, built by Zarate Salmeron in 1626, was burned several times during the 1600s. Ultimately, it was abandoned in 1632.

Alonso Benavides (1630), wrote a similar plea to the king in 1630. In his memoria, he claims that *eighty thousand* Pueblos had converted. At this pace, the friars would have to have converted twenty-five Pueblos every day for five years straight! Both memorials are highly fictionalized and inform the king of the presence of great mining wealth and tribute-rich populations. Benavides also claims to have met a mystical nun in Spain who traveled back and forth between New Mexico and had witnessed the baptisms, among other things. In 1638, his superior, Custodia Juan de Prada (who never visited New Mexico), estimated that there were less than forty thousand Indian souls in the region.

By this time the struggles between secular leaders and missionaries over a dwindling supply of Indian labor had led to calls for the abandonment of the region. In 1632, the residents of Hawikuh killed the resident priest and demolished the church after being forced to attend mass (Ivey 1998: 127). In 1623, the Jemez began the first of their rebellions against their missionaries. After 1623, the area was never fully pacified. (Scholes 1936: 147). A repopulated Acoma remained a stronghold for Keres people resisting missionization (Hodge 1945: 72). Fray Benavides in his memoria recounts several incidents of defiance. Preaching at the Tompiro pueblo of Chilili, Fray Benavides was challenged by a Cacique:

> [He said], "You Christians are crazy. You desire and pretend that this pueblo shall also be crazy." I asked him in what respect we were crazy. He had been, no doubt, in some Christian pueblo during Holy Week when they were flagellating themselves in procession, and thus he answered me: "How are you crazy? You go through the streets in groups flagellating yourselves, and it is not well that the people of this pueblo should commit such madness spilling their own blood by scourging themselves." When he saw that I laughed, as did those around me. he rushed out of the Pueblo, saying that he did not wish to become crazy. (Hodge 1945: 66).

While attempting to convert the son of a Cacique at Picuris pueblo, a Fray Letrado was beaten and killed by villagers who "rose in rebellion and attacked him in a body, smashing his head with their clubs in order to prevent him from preaching the word of God any longer" (Hodge 1945: 78). At Zuni, Hopi, and Acoma, similar incidents occurred (Hodge 1945: 69–78). After "healing" a blind child at Hopi, a missionary was poisoned: "the envy of the priests of the idols was so rabid that they again incited the people to revolt . . . [and] killed him with poison" (77).

At Zuni, a Fray Arvide was killed by "an old sorcerer" who "struck the blessed father with such a blow that felled him" (Hodge 1945: 78) At Jemez, Fray Zárate Salmerón's church and mission were burned after attempting to relocate the Towas to the mission, "with the result that it was totally abandoned and all of the Indians returned their former mountains, and most of them scattered over other regions" (Hodge 1945: 70).

Clearly, by the mid–seventeenth century, missionary efforts among the Pueblos were in full swing. Following the tactics of conversion developed by Franciscan missionaries in Nueva Vizcaya, few, if any of the priests were capable of communicating the complex theological principles of Catholicism to the Pueblos in any detail (Knaut 1994: 180). For the Franciscans of the 1600s, "conversion" was demonstrated through participation in highly routinized, ritual practices; baptism required neither a clear awareness on the part of the participants as to the significance of the act, nor any intellectual or spiritual transformation. Beyond the veneration of the cross and respect for the clergy, there was no attempt made to indoctrinate the Pueblos in the intricate theological principles that characterized European Catholicism. Recognizing these limitations, large wooden placards indicated (through paintings) the presence of heaven and hell, Jesus, Mary, angels, and the devil (Phelan 1980). Given these limitations and the use of Native interpreters, it is unlikely that in these early years the Pueblos conceived of Catholicism by the same terms as the Spaniards.

Church and State Conflicts in the 1600s

Both civil and ecclesiastical officials were accused of forcing the Pueblos to labor without pay, but the Pueblos appear to have made concessions to the governors and encomenderos in exchange for increased religious freedom. Governor Juan de Eulate's (1618–1625) was accused by Custodia Parea of refusing military escort into the Pueblos, selling slaves to the mining areas in Zacatecas, hindering the enforcement of religious instruction, and declining to cooperate with the friars in suppressing Puebloan ceremonies (Scholes 1936: 148).[47] In later reports, Eulate was implicated in the revolt of the Jemez Pueblos in 1623. Arrested in 1627 for bringing captives from New Mexico to Mexico City, Eulate's successors would engage in similar behaviors. (Scholes 1936: 160, 164).

Governor Baeza (1634–1637) increased tribute levels in exchange for relaxing the enforcement of religious instruction (Scholes 1936: 287).[48] Governor Luís de Rosas (1637–1641) established a textile factory and workshop in the capitol at Santa Fe using forced Indian labor. The custodia later accused him of allowing the Pecos to practice their "idolatry" in exchange for increased tribute levels. Rosas was similarly charged with staging raids against Jumano (Apache and Navajo) communities in order to collect slaves for his workshop and for the Zacatecas region (Scholes 1936: 301). During Rosas's tenure, the Pueblos at Taos rebelled and killed their priest. Demanding retribution from the governor, an open conflict ensued between the governor and the friars; Rosas jailed two of the missionaries, and the priests refused to enter the Tewa pueblos for over a year (Scholes 1936: 321–322).

Apparently, colonists loyal to the governor sacked the conventos at Sandía, Cuarac, and Soccoro (Scholes 1936: 323) Later, Rosas was murdered by another colonist. Meanwhile, slave raids among the Apaches led to counterattacks. In a letter to the king, Fray Andrés Juárez argued that expeditions had been staged throughout the 1600s for the purpose of capturing and selling Apache slaves to mining communities in Zacatecas and that the Pueblos had returned to their traditional religion. Aided by Apache allies, Jemez, Cochiti, Isleta, San Felipe, and Alameda rebelled in the 1640s (Scholes 1936: 99, 105). Rebellions similarly occurred at Hopi after two Hopi men were whipped and tarred by the missionary in residence for performing Catholic ceremonies in the church (Scholes 1942a: 13–14).

The administrative changes of the latter half of the seventeenth century brought both increased freedom for the Pueblos and heightened tensions between the colonial factions. Missionary levels had fallen in the early 1600s, leaving only forty-six Franciscans in New Mexico by 1654.[49] Most significant were the activities of Governor López de Mendízabal. During his tenure, López attempted several reforms that aggravated the relationships between civil and church officials. It is worth noting that López was reportedly descended from conversos (Scholes 1942a: 98–106, 146, 149–156). Later accused by the missionaries of secretly practicing Judaism, López (perhaps logically) appears to have held very liberal views regarding the cultural significance of Puebloan ceremonies.

In his *audiencia*, López was vilified by the Franciscans for instituting a series of reforms among the Pueblos. Tribute amounts were temporarily lowered, missionaries and colonists were forced to pay the Pueblos for work, and encomiendas and government appointments were reapportioned (Scholes 1942a: 25, 40). Reversing the policies of the Franciscans outlawing the public performance of katchina ceremonies at all pueblos, López invited Tewa dancers to perform in the plaza at Santa Fe. After viewing the dances, López was heard to say, "These scoundrel friars say that this is evil; it is not evil, but very good; and but for the fact that I am governor I would go out and dance myself" (Scholes 1942a: 61–62). Refusing to enforce Puebloan participation in religious services and forbidding the friars from administering public punishments to resistant Pueblos, López investigated complaints from Natives at several villages. He removed or publicly punished priests accused of sexual misconduct and relocated encomenderos living within the Pueblos.[50] Alarmed at the loss of authority, Franciscan authorities reported López to Inquisition officers in Mexico City.[51] Lamenting the return of the Pueblos to their traditional religions, the custodia of New Mexico reported that

> the ministers [of the missions] have suffered . . . great persecution and dishonor, and the Indians are totally lost, without faith, without law, and without devotion to the Church; they neither respect nor obey their ministers, and it makes one weep to see that in such a short time they have lost and forgotten what they have been taught all these years. . . . In the years that I have unworthily served as guardian at this convent, I have not seen the said governor or any minister of justice punish any fornicator, idolator or sorcerer in this pueblo. (Scholes 1942a: 89–90)

In response to these complaints, Inquisition officials appointed a new custodia (Fray Alonso de Posada) to accompany incoming Governor Diego de Peñalosa. Reversing López's edicts, Posada's first actions included the confiscation and destruction of Katchina masks and other ritual objects within the pueblos (Scholes 1942a: 98). Following the custodia's recommendations, both López and his wife were imprisoned by Peñalosa and transported to the Inquisition prison in Mexico City.[52]

The Ethnogenesis of the Pan-Indian Movement in New Mexico: Pre-Revolt Movements in the Seventeenth Century

Custodia Posada's crackdown on Puebloan ceremonies coincided with several events that contributed to the emergence of the pan-Puebloan movement of the 1680s. An extended drought between

1666 and 1670 increased the strain on the agricultural production of the pueblos.[53] Despite the starvation of the Pueblos, encomenderos and civil and ecclesiastical officials continued to exercise their rights to appropriate Puebloan food. Notwithstanding López's brief period of reforms, tribute levels among the pueblos had steadily increased from one fanega of corn per household biannually in 1610, to four fanegas (about four hundred pounds) per person per year after 1643 (Gutiérrez 1991: 117).

Newly appointed governor Treviño (1675–1677) reinforced Posada's crackdown on Puebloan ceremonies and ordered kivas to be filled and religious objects confiscated and burned (Hackett and Shelby 1942: 289–290). These actions can be interpreted in several ways. Treviño may have wished to gain the approval of the missionaries through a show of force, or the crackdown may have signaled an attempt to regain authority that had already been lost. The historical documents indicate that in the years between 1640 and 1650, the missionaries steadily lost control of the missions in the southern Tiwa and Piro provinces, as well as at Taos, Jemez, and Zuni. In separate incidents, the conventos and churches in each of these regions had been burned and the missionaries killed by alliances of Puebloan and Navajo-Apache warriors (Espinosa 1991: 31). In 1644, an alliance between Jemez and Apache warriors was discovered, and the leaders were executed. A similar pact between Isletan Alamedan, Cochiti, San Felipe, Jemez, and Apache warriors had been discovered and crushed in 1649 (Hackett and Shelby 1942: 245–246, 266, 299). In 1650, the southern Tiwa Pueblos of Alameda and Sandia formed an alliance with Apache allies and attempted "to destroy the whole kingdom, and to be left in freedom as in ancient times, living like their ancestors" (Hackett 1942: 299). The revolt, foiled by Christianized Indians, was followed by a similar attempt a year later. A second movement was attempted in the southern Tiwa and Piro pueblos in 1668, but the plot was discovered, and the leader, a Cacique from Abó named Estéban Clemente, was executed (Hackett and Shelby 1942: 245–246, 266, 299).

In each of these cases, Christian Indians (some of whom were likely *genízaros*, Indigenous captives who served as slaves) living at the pueblos had revealed the plots. Those suspected of instigating the rebellions (often Puebloan religious leaders) were executed or sold into slavery. The social costs of remaining loyal to the religion, culture, economic structure, and ideology of traditional Puebloan social life were high. Enslavement, persecution, beatings, labor, and tribute could all be visited upon those Pueblos who still lived in close association with their Spanish overlords. Clearly by this time the Pueblos were also aware of the risks associated with living near partially Hispanicized Pueblos, and Spanish policies were specifically designed to exploit these divisions (Jones 1966: 3–33). Pueblo children, abducted from their parents or orphaned because of their Spanish blood, frequently served as intermediaries between the two worlds.[54] Because of their positions, many of those described as "half-breeds," coyotes, or *lobos* were capable of interpreting, communicating, and navigating between the dichotomized worlds of the Pueblos and the colonists.[55]

For those Pueblos who wished to retain their social identities as members of distinct linguistic and cultural communities, the cost of resistance was high as well. Faced with forcible relocation and cohabitation with missionaries, it is clear that many Pueblos chose to sacrifice their homelands, sacred landscapes, and lifestyles in order to maintain some degree of control over their own traditions. Social identities, alliances, and affiliations with competing social identities (Pueblos vs. Christians vs. Apaches and Navajos) were constantly reformulated and reorganized during these tumultuous years. Estéban Clemente had apparently feigned allegiance with the missionaries in order to gain trust and access to information about the colonists. After his execution, his home was found to have served as a refuge for sacred objects during the years of Treviño's repressive actions (Hackett and Shelby 1942: 300). Ambiguous social identities and social actions appear to have been a prevalent feature of social interactions between the Pueblos and the colonists. After the successful rebellion of 1680, the missionaries noted, "The Indians who have done the greatest harm are those who have been most favored by the religious and who are most intelligent" (Hackett and Shelby 1942: 59).

The historical relations and tensions between Christian and "traditional" Indians appear to have led to a series of coordinated attacks (or "rescues") among missionized Pueblos. Confederated Apache, Navajo, and Pueblo raids against conventos and "converted" Pueblos increased in the 1670s. In 1673, the convento at Zuni was sacked, two hundred Zunis were captured, and the friar was killed. Attributed to a fictitious Apache raid, the event seems to have been motivated by a desire to recapture the loyalties of Natives loyal to the Spanish (Forbes 1960: 150). Similar attempts at maintaining and enforcing social distances between Spaniards and Pueblos were common throughout New Mexico. In a letter to the king, Fray Ayeta stated that the pueblos of Los Humanas, Senecú, Abó, Cuarac, Chilili, and Las Salinas had been steadily depopulated during the late 1600s (Schroeder 1979: 241). Ayeta attributed the abandonments to "the barbarous rabble of Apaches and other confederated heathen nations" (Hackett 1937: 297). Other events raise questions about the identities and motives of the participants in the raids as well as the causes of the abandonments in the Tompiro and Piro districts. At both Abó and Senecú, Franciscan priests had been killed and the conventos desecrated, but throughout the seventeenth century, missionary efforts among the Apaches and Navajos were largely nonexistent. The desecration of religious objects and the attacks on the priests are more consistent with the attacks of the Pueblos during the Revolt. The chief source of Apache and Navajo anger toward the colonists had been the slave-raiding activities of the governors and encomenderos.

It seems likely that by the late 1670s the seeds of a pan-Indian ethnic movement had begun to develop. The "confederated heathen nations" likely consisted of traditionalist Pueblos (interested in the restoration of religious authority) as well as their Apache and Navajo allies. Other reports indicate the fluidity of social relations between the Pueblos and their Athapaskan neighbors. One missionary recounted that, "upon the slightest occasion of annoyance with the soldiers . . . some of the baptized Indians, fleeing from their pueblo, have gone over to the heathen, believing that they enjoy greater happiness with them" (Hackett 1937: 298, 111, 106–114). Given the relatively small numbers of missionaries during the seventeenth century (forty-six in 1656), the clandestine performances of rituals among the Pueblos could hardly have been effectively monitored (Espinosa 1988: 24).

The breaking point between the colonists and the Pueblos was finally reached in the summer of 1680. Governor Treviño ordered the arrest of forty-seven Pueblo religious leaders, three of whom were hanged at Nambé, San Felipe, and Jémez (Hackett and Shelby 1942: 300; Guitiérrez 1991: 131). The remaining Caciques and leaders were publicly beaten and threatened to be sold into slavery. In retaliation, Santa Fe was surrounded by Pueblo warriors, and Treviños quarters were stormed. The Pueblos demanded the release of the captives; after a tense standoff, Treviño capitulated. One of the captives, a Tewa religious leader from San Juan, fled to the pueblo of Taos and began to organize the successful rebellion of 1680.

Social Violence, Mobility, and Disease: A Colonial Mythology in Need of Critical Reanalysis

The documentary history of the entradas is quite startling. A close reading of primary sources demonstrates the degree to which the mythology of colonization needs to be reexamined. This was an absolutely brutal period for the Pueblos and their southern neighbors—and disease seems to have had nothing to do with it. Until this point, no mention had been made of any of the effects of epidemics that were supposed to have raged through the Southwest prior to colonial contacts. What we do see is the devastating effects of aggression as a tool of control—always legally sanctioned and framed in the documents as either necessary to secure resources or preceded by the reading of the requirement. Similarly, slave raiding, the confiscation of all available food and provisions, and the large movements of Pueblo people away from Spanish colonists are undeniable components of this colonial history. One must ask, Why has this history been largely ignored by archaeologists?

These events have nothing to do with any Black Legend invocations. They were not written by the enemies of Spain, nor are they the product of "anti-Hispanic sentiment." They were written by the participants themselves. Their conspicuous absence from much of the secondary historical literature and the emphasis of that literature on "disease" as the primary agent of colonial destruction give strong evidence supporting the notion that Spanish aggression and wanton destruction were important, even celebrated activities in these encounters, and that many historians' discomfort with (and archaeologists' ignorance of) these readings has led to an overly sanitized version of conquest. Perhaps five hundred years after these events, we can confront this issue in an honest manner and reread these accounts from an Indian perspective. The fact that this has not yet happened reveals the strength and endurance of the colonial mythology of conquest, and the need for a thoughtful reexamination of not only the documents but the interpretation of archaeological sites in the region.

The social consequences of contact are in similar need of reexamination. The primary source materials are revealing on several levels. First, the documents consistently support the correlation of social violence and abandonments *during the entire colonial period*, and the abandonments themselves demonstrate a single but very important element of Puebloan resistance. We know, for example, that during the beginning of the seventeenth century, many of the Pueblos south of Santa Fe (Piro and southern Tiwa) were gradually depopulated throughout the colonial period (Schroeder 1979a, 1979b). Although these abandonments are attributed to Apache and Navajo raids in colonial documents, other factors clearly led to the abandonments. The failed rebellion of Estéban Clemente appears to have led directly to regional population shifts away from the mission centers.

Where the participants of these movements fled to is unclear. They may have joined up with Apache and Navajo allies, or they may have fled north and resettled in communities where the colonists appear to have lost control. Possible relocation areas could have been the northern Tiwa villages of Taos or Picuris, or other, as yet unidentified settlements could have housed these refugees. This theory might explain some of the events of the Revolt. We know, for instance, that many of the southern Pueblos were apparently not notified of Po'pay's rebellion. It is possible that after Clemente's revolt, the only Pueblos still in the region were those who remained loyal to the colonists. Still, more research needs to be done in the more remote mountainous villages north of Santa Fe. Others may have fled to the Jemez, Zuni, or Hopi territories.

From a social and cultural standpoint, the historical experiences of persecution, violence, and ethnic subordination have an important symbolic value. Persecution provides an important ingredient to pan-ethnic movements. These processes facilitate both sympathy and empathy between subordinate group members. More importantly, these collective historical experiences emphasize a sense of shared fate and actually help create an awareness of "community" among subject peoples. In this sense, persecution enhances the status positions of victims and creates opportunities leadership. As we shall see in the following chapter, Po'pay as an ethnic entrepreneur was arguably one such figure. Ethnic entrepreneurs are able to rally people around the symbols of their oppression. Ethnic mythologies are thus enhanced by the symbolism of the dominant group and explain who the "we" are in relation to outsiders. Ethnic movements generate their own ritual and sacred practices and redeploy the symbols of oppression (in this case, sacred Spanish objects) as emblems of resistance.

The following chapter addresses the subjects of abandonment and refugee movements using archaeological data from the Jemez region. The Jemez Pueblos were fairly successful in maintaining cultural practices through armed rebellion, the destruction of colonial centers, and population shifts, and the Jemez Mountains were the site of several of the more well-known refugee Pueblos. While this region represents only a single possible area for the relocation of rebellious groups, many of the larger historic period pueblos in the area have been identified by several different archaeological projects during the past forty years. Although population figures are difficult to assess without excavation, this archaeological information provides evidence supporting the reoccupation of ancestral Jemez villages in one of the more remote areas of northern New Mexico.

CHAPTER 6

"Seek and You Shall Find"

Mobility as Social Strategy: Documenting Evidence of Contact and Revolt Period Settlements

The Statements of closet historians that that the Spanish enslaved the pueblos, or any other Indians in New Mexico, that they forced them to choose between Christianity and death; that they made them work in the mines and the like are all entirely untrue. . . . Yet the mere presence of the Strangers in their land was enough to stir the jealous nature of the Indians; and in 1680 a murderous and causeless plot broke out in the red Pueblo rebellion. . . . On that bitter 10th of August, 1680, over four hundred Spaniards were assassinated—including twenty one of the gentle missionaries who, unarmed and alone, had scattered over the wilderness that they might save the souls and teach the minds of the savages.

Charles Lummis, *The Spanish Pioneers* (1909)

This revolution [the Revolt of 1680] was fought for precisely the same reasons that the revolution of 1776 was fought—to regain freedom from tyranny, persecution and unjust taxation. . . . It was a fight for freedom that has never been widely recognized as such, nor accorded the attention by historians that it might have had . . . but it is part and parcel of our common heritage as Americans and it must be made part of our common consciousness.

Alfonso Ortiz, cited in *Indian Uprising of the Rio Grande* (Folsom 1973)

The words of early Borderlands historian Charles Lummis and Alfonso Ortiz, a Native American anthropologist raised at the Tewa Pueblo of San Juan, reveal the degree to which historical narratives of contact, conquest, and rebellion are part of ongoing political processes. Ethnohistories are continually refashioned, reshaped, and redeployed to support the interests and perspectives of various ethnic groups. In the preceding chapter, I presented historical documents supporting the correlation between the use of social violence by colonists and missionaries and regional abandonments during the entradas of the sixteenth century. Following the model of ethnic conflicts outlined earlier, I suggested that mobility, abandonment, and relocation are characteristic responses of colonized populations to the processes of pan-ethnic ascription and colonial violence. This chapter addresses a central issue in the reformulation of conquest social dynamics and the use of mobility as a social strategy. It addresses a question central to the reanalysis of postcontact Pueblo histories: if the Pueblos attempted to maintain social distance and preserve cultural differences through regional and local abandonments, is there archaeological evidence supporting these movements?

I attempt to answer this question using archaeological and settlement data from one of the more remote and poorly controlled regions of the Pueblo region. As demonstrated in the previous chapter, the Jemez, northern Tewa, Hopi, Zuni, and Keres provinces were the sites of ongoing Puebloan resistance throughout the seventeenth century. Unfortunately, regional studies of Puebloan population movements during the seventeenth century have been difficult to re-create using archaeological data. For reasons outlined earlier, synthetic analyses of nonmission historical Puebloan sites have been generally hindered by a lack of archaeological research (Ayers 1995; Schroeder 1979a, 1979b). Disease narratives have played a large role in creating this information gap. If abandonments are caused by epidemics, there is simply no reason to seek evidence supporting further settlement.

It must be mentioned that few Pueblos are eager to have archaeologists sink excavation test trenches in the center of occupied villages. Still, many sites that are normally associated with earlier occupations contain historic and postcontact period ceramics. Because of the relative similarity of the most commonly used diagnostic pottery (glazewares), I would argue that much of this evidence at other sites in the region has been either missed ignored. This is an important arena of future research.

Recent research by Jeremy Kulishek in the analysis of Jemez region field houses is most intriguing. Kulishek (2005) examined thirty field houses in the remote mountains surrounding the Towa village of Jemez. He has found evidence supporting the intensification of field house usage from 1525 to 1650. This runs counter to many researchers' expectations regarding population declines in the region and actually supports the notion that populations expanded in this region, in particular during both the entrada (1540–1600) and pre-Revolt periods. He argues convincingly that warfare, migration, and shifts in ethnic identity—not disease—were of central importance in determining population shifts. This work generally supports my own research. As pan-Indian (pan-ethnic) movements gain momentum, the boundaries that separate different kinds of Indians become less important. The common experience of subjugation and stigmatization facilitates the free movement of people within the general category of "Indian." Whereas vertical movement (to a higher ethnic status) is impossible, mobility within these lower-status positions opens up. Using a different data set, archaeologist Matt Leibman (2006), working in conjunction with the Pueblo of Jemez, has further substantiated the importance of the Revolt in the generation of ethnic consciousness among the Towa.

The archaeology of the Jemez forest region has attracted significant attention from archaeologists in recent years. As many of the Revolt period sites are located in the relatively isolated Jemez National Forest, Forest Service archaeologists have been able to map and collect artifact samples from many of these large, multistoried ancestral Jemez villages. These data, collected from fieldwork and Forest Service reports, provide evidence supporting the periodic reoccupation of Jemez sites during the sixteenth and seventeenth centuries. Using the Jemez data as a sample set of Puebloan historic movements, it is hoped that this work will encourage similar contextual studies of historic Puebloan migrations in other regions. In chapter 7, I examine the archaeological evidence collected from one of the more well-known Revolt period sites, Old Cochiti. This analysis provides evidence supporting the development and dissolution of the pan-Puebloan movement during the period preceding the recolonization of Don Diego de Vargas in 1693. Before discussing the archaeological information, a brief review of the chronology and social consequences of the Revolt will be presented.

The Development of Pan-Puebloan Consciousness and the Revolt of 1680

The transformation of Puebloan social, economic, and religious activities by colonial officials led to dramatic changes in the demographic structure of Puebloan communities. Further disruption was caused by the colonial policies of reducción. The removal and resettlement of Puebloan communities into larger, more easily monitored villages altered and reduced the subsistence bases of smaller villages. Agricultural production was curtailed by the forced abandonment of small, widely scattered fields and field houses at the regional level. In limiting the mobility of the Pueblos, the delicate balance between the production of agricultural resources and the supplementation of the Puebloan diet with communal hunting and gathering were severely curtailed.

During the colonial period, surpluses gathered to ensure the survival of communities through the well-known cycles of drought and abundance were confiscated to serve the needs of the colonial population. The resettlement of formerly independent villages into larger, more easily managed communities, coupled with the food shortages created by tribute demands, accelerated the deterioration of Puebloan community health. While disease played an important role in the depopulation of Puebloan villages, the strains created by the disruption of the agricultural cycles, carefully managed by Puebloan religious leaders, severely compromised the health of Puebloan peoples. Each of these factors caused the Pueblos to consider the most drastic of responses: armed rebellion and the forcible removal of Spanish colonists.

The events surrounding the Revolt are fairly straightforward. Following a series of unusually dry years and severe famines in the early 1670s, Pueblo religious leaders were accused of bringing the misfortune on the land through the practice of witchcraft (Hackett and Shelby 1942: 289). In retaliation, then governor Treviño ordered that forty-seven Pueblo religious leaders be imprisoned and publicly beaten; four of the Pueblo leaders were eventually slain by Spanish authorities. In retaliation, Pueblo warriors stormed Treviño's quarters and threatened him with death unless he released the remaining leaders. After a tense standoff, Treviño capitulated, but colonial officials continued to single out a single religious leader, a man from San Juan named Po'pay (Hammond 1942: xxii). Eventually Po'pay was forced to leave San Juan and was invited to live in the northernmost pueblo of Taos (Hackett 1942: 242). Fearing the total destruction of their cultures, religions, and communities, the Pueblos determined that the Spanish must be forced out.

Po'pay enlisted the aid of several religious leaders from other Pueblos. Among them were several of the Spaniards' most trusted, Spanish-speaking intermediaries. At this meeting were the coyotes Alonso Catití from Santo Domingo, Domingo Romero from Tesuque,

Nicholas Jonva of San Ildefonso, and Domingo Naranjo from Santa Clara (Hackett 1942: 246; Gutiérrez 1991: 132; Sando 1998: 195).[1] Joining the rebellion were Francisco El Ollita, the war chief of San Ildefonso; the Keres leader Antonio Malacate; Luis Tupatú from Picurís; as well as war chiefs from Taos, San Lorenzo, Jemez, and Pecos (Hackett 1942: 246). Fearing the detection of their activities, they held secret meetings during the saint's day festivals. Determined to attack the Spaniards when supplies of gunpowder, military equipment, and food were low, the leaders timed the rebellion during the months preceding the arrival of the mission supply caravan (Hackett 1942: 234–235).

On the night of August 9, 1680, two young runners named Omtua and Catua were dispatched from the Tewa pueblo of Tesuque carrying the instructions for the rebellion. As they visited individual Pueblos, each morning the runners were to untie a single knot of a *magey* fiber cord signaling the date of the attack. On the final morning, August 11, the individual pueblos were to take up arms against each of the Spanish communities in their area (Hackett 1942: 246). Unfortunately, when the runners reached the Pueblos of Pecos, San Cristobal, San Marcos, and La Cienega, their message was intercepted by Natives sympathetic to the Spanish colonists. Fearing the rebellion, newly appointed governor Don Antonio de Otermín dispatched riders to warn the colonists in the surrounding area of the attack, but it was too late. Attacks were staged at individual pueblos, and estimates of casualties among the colonists reached 380. More than half of the forty Franciscan missionaries were slain by their parishioners (Hackett 1942: 107, 161).

Otermín collected the remaining colonists at the plaza in Santa Fe where he was able to hold out until the water supply feeding the compound was diverted. Suffering from thirst and exhaustion, the governor staged a counterattack, and on August 20, he captured and executed forty-seven warriors. Overwhelmed, Otermín was allowed to abandon the area unmolested with whatever provisions he and his followers could carry. It is noteworthy that the genízaro settlement of San Miguel, containing many "Mexican Indians" (likely Tlaxcallan), was targeted by the Pueblos as well (Hackett 1942: 13–14,171).

After a series of failed counterattacks, Otermín fled south to El Paso del Norte with approximately 1,500 Indian servants and allies. He remained at El Paso for over a year, gathering his forces. Ordered by the viceroy to retake the colony, Otermín set out for New Mexico in November 1681. With 146 soldiers and 116 Pueblo allies, Otermín's party found the region completely abandoned. The missions and religious materials had been burned, the bells had been dismantled, and sacred images were hacked to pieces (Hackett 1942: cxxvi). Moving north, Otermín learned that the natives of Alameda, Puaray, and Sandia had fled to the Sandia Mountains; people from San Felipe, Santo Domingo, and Cochiti had entrenched themselves at Old Cochiti; and Indians from Jemez, Santa Ana, and Zia had fled into the villages of the Jemez Mountains.

Reaching Cochiti Mesa, Otermín's captain Juan Domínguez de Mendoza attempted to retake the settlement. Finding the mesa fortified with over one thousand warriors from many different villages, Mendoza negotiated a temporary peace but was unable to force the abandonment of the any of the Revolt settlements (Hammond 1942: clxxi). Fearing for his own safety, Mendoza retreated south with the remainder of his party. Word of the Pueblo success apparently traveled quickly through the frontier. Emboldened by the Pueblo victory, Manso, Jano, Suma, Taboso, Julime, Concha, and Pima Indians attempted to overthrow the colonists in their regions (Naylor and Polzer 1986: 483–718). With their resources divided, former governor Posada and Governor Jironza led failed attempts to reconquer New Mexico in 1686, 1689, and 1691. Capturing several slaves and executing prisoners, these expeditions only reinforced the resolve of the Pueblos to remain in their revolt strongholds.

The Pueblos were to remain free of colonial domination until 1692 when Don Diego de Vargas reentered the region with a party of colonists, soldiers, and Indian allies. (Espinosa 1988: 37; Kessell and Hendricks 1992: 22–25; Naylor and Polzer 1986: 506–547). The presence of Indian allies among de Vargas's troops appears to have signaled the partial dissolution of the pan-Puebloan movement. As discussed later, after achieving the immediate objectives of the Revolt, the rhetoric of resistance and the call for a return to ancestral religious practices and customs appears to have reinvigorated local identities. The following section explains how these events closely parallel the general pattern of ethnic movements introduced in chapter 3.

Analysis and Implications: The Development and Demise of the Pan-Puebloan Movement

Following the regional rebellion, the Pueblos had achieved one of the most successful Indigenous revolts in the history of the Americas. The pan-ethnic ascription that relegated the Pueblos to a subordinate social category appears to have actually created the conditions in which rebellion was inevitable. Despite the fact that different Indigenous peoples had communities that they themselves considered distinct culturally, located within their own historical traditions, the Spanish had defined and treated the Pueblos as a single subordinate class of people. The collective resentment generated by the long history of subjugation, violence, and the repression of Puebloan religions was revealed in the testimony of an eighty-year-old Tiwa man interviewed following the Rebellion. When asked why the Indians had rebelled, he stated that

> the resentment which all of the Indians have in their hearts has been so strong, from the time this kingdom was discovered, because the religious and the Spaniards took away their idols, and forbade their sorceries and idolatries: that they have inherited successively from the old men the things pertaining to their ancient customs. And that he has heard this resentment spoken of since he was of an age to understand. (Hackett and Shelby 1942: 61)

Following a pattern typical of ethnic conflicts, the practices of subordination often lead to unintended consequences for the dominant group. While some individuals attempted to integrate themselves into the colonial system through the adoption of Spanish cultural practices, eventually these attempts were thwarted by the diacritics of the dominant society. This pattern of behavior is frequently observed in ethnic conflicts. While some communities may attempt to remove themselves from the social system through migration or refugee movements, other groups attempt to coexist with the dominant group.

As Horowitz (1985) documents in his comparative study of ethnic conflicts, eventually the common experience of discrimination and marginalization motivates collective social behaviors or pan-ethnic movements that attempt to remove the source of inequality and restore equilibrium and group autonomy (1985: 3–55). These movements often adopt the shared diacritics of their subordinate group and redeploy them as symbols of a common identity. Segregation between the groups accelerates, and languages and religious practices are cleansed. Group members also characteristically seek a return to orthodox practices that existed prior to contact with the dominant group.

Po'pay's call to arms was accompanied by a similar set of invocations. During the rebellions in 1623 at Jemez and 1632 at Zuni, in 1639 at Taos, as well as the unsuccessful regional rebellions in the Tiwa provinces in 1650, religious leaders were executed or sold into slavery, aggravating a growing discontent among the Pueblos (Knaut 1994: 98–99; 169). The evidence for the assertion of orthodox practices is provided by the testimony of and oral histories of Pueblos following the Revolt. Po'pay urged the renouncement of Spanish beliefs and customs, calling for ritual purification and the reinstatement of traditional ceremonies. Po'pay ordered "that all of the pueblos break up and burn the images of the Holy Christ, the Virgin Mary and the other saints, the crosses, and everything pertaining to Christianity and that they burn the temples, break up the bells and separate from the wives whom God had given them in marriage" (Hackett and Shelby 1942: 247–248).

The returning Spaniards found religious icons with arms hacked off, burning in piles reminiscent of the public destruction of Puebloan ceremonial objects:

> As blind fiends of the devil, they set fire to the holy temples and images, mocking them with their dances and making trophies of the priestly vestments and other things belonging to divine worship. Their hatred and barbarous ferocity went to such extremes that in the pueblo of Sandia images of saints were found among excrement, two chalices were found concealed in a basket of manure and there was a carved crucifix with the paint and varnish taken off by lashes. (Hackett and Shelby 1942: 177–178)

Ritual purification was necessary in order to reverse the processes of integration into Spanish religious life. "In order to take away their baptismal names, the water and the holy oils, they were to plunge into the rivers and wash themselves with amole, which is a root native to the country washing even their clothing with the understanding that there would thus be taken from them the character of the holy sacraments" (Hackett and Shelby 1942: 247–248).

Finally, Po'pay called for the restoration of Puebloan cultural practices, the rebuilding of the kivas, and the desecration of Catholic religious icons. The Pueblos were called on to "discard the names given to them in holy baptism and call themselves whatever they liked. . . . They were ordered likewise not to teach the Castilian language in any pueblo and to burn the seeds which the Spaniards sowed and to plant only maize and beans, which were the crops of their ancestors" (Hackett and Shelby 1942: 235). The return to the ways of their ancestors would allow the Pueblos to

> harvest a great deal of Maize, many beans, a great abundance of cotton calabashes and very large watermelons and cantaloupes; and that they could erect their houses and enjoy abundant wealth and leisure . . . and that they thereby returned to the state of their antiquity, as when they came from the lake of Copala . . .

because the God of the Spaniards was worth nothing and theirs was very strong, the Spaniard's God being rotten wood. (Hackett and Shelby 1942: 247–248)

Po'pay's call for a return to orthodox practices led to the settlement of a series of pan-Puebloan communities in the years following the Revolt. Members of San Cristobal and San Lazaro established Tewa Village on First Mesa at Hopi (Dozier 1954). La Alameda was founded on Cerro Colorado by people from Zia and Santa Ana (Kessell and Hendricks 1992: 431). And as mentioned previously, on Cochiti Mesa, Cochiti people hosted members of San Marcos and San Felipe (Kessell and Hendricks 1992: 515). By 1694, members of San Ildefonso, Santa Clara, San Juan, Tesuque, Cuyamunque, Pojoaque, San Lázaro, San Cristóbal, and Jacona pueblos had established a stronghold on the summit of Black Mesa (Kessell, Hendricks and Dodge 1998: 116). In the Jemez Mountains, Jemez people established Patokwa and Astialakwa on a large mesa in the Jemez region (Bloom and Mitchell 1938). Several other ancestral Jemez villages were also reoccupied during the period (Elliot 1982).

Ethnographic documents and oral histories can offer clues as to how the pan-Puebloan movement was activated. At the local level, the establishment of new settlements implies a restructuring of communal identities, kin and clan affiliations, and communal relations. Choosing to resettle involves a deliberate decision to increase social interaction. With these interactions comes an increased likelihood of social friction. It is instructive, then, that the settlement of these communities closely parallels linguistic affiliations. At Cerro Colorado, Zia and Santa Ana peoples spoke Keres dialects. At Black Mesa were speakers of northern Tewa dialects. Tanoan speakers from San Lazaro and San Cristobal migrated to First Mesa at Hopi.

An important strategy of resistance mentioned by Dozier, Ortiz, and other Pueblo scholars was the protection and use of language as a means of insulating Pueblo communities from unwanted outside influences. Following the policies of Franciscan society, few, if any, of the Franciscans made efforts to learn the multitude of Puebloan languages (Phelan 1970: 86–91). Missionaries interested in maintaining control over the Pueblos enlisted the help of translators in communicating Catholic doctrine. Though intended to consolidate authority, the failure of the Spaniards to create an Indian clergy led to syncretic interpretations of Christianity. This allowed the Pueblos to interpret Catholic doctrines on their own terms. Heavily laden with the thick description of familial relations, cosmology, and religion, language helped affirm communal solidarity and reestablish local identities.

Still, Po'pay's call for a return to fundamentalist practices and his attempts to assert regional authority had mixed results. While the regional movement was successful, few communities were willing to accede authority to a centralized leadership (Twitchell 1914: 171). As the threat of Spanish domination was removed, the regional system lost momentum. Similar to other pan-ethnic movements, the call for a return to a new fundamentalism invokes different versions and notions of orthodox practices. The successful removal of the colonists helped motivate new objectives for individual communities. Ultimately, no group was prepared to accede authority to a single leader, and in the end, local identities resumed as the ascendant forms of social organization. In many ways, the Revolt succeeded in changing the balance of power between the colonists and Pueblos.

With the reentry of a new generation of colonists in 1693, the terms and conditions of interaction between the groups changed. A higher degree of religious freedom was established, and cultural practices were restored. Encomiendas and the tribute system were also dissolved after the Revolt (Kessell 1989: 132–133). Once activated, the pan-Puebloan identity remained a significant obstacle to missionization and forced labor. When a new call for rebellion was issued in 1696, Tewa, Jemez, and Tano Pueblos resumed their attacks against the missionaries (Espinosa 1942: 296–304). Between 1696 and 1706, the Jemez people abandoned their homeland and sought refuge with Zuni, Acoma, Laguna, Hopi, and Navajo allies (Carlson 1965; Hester 1962).

As the documents presented in the previous sections demonstrate, the Jemez region was one of the most important centers of Puebloan resistance. Five different missions were successively built and destroyed by the Jemez (Reiter 1938: 41). During governors Otermín's (1681) and Jironzas's (1688) failed attempts at reconquest, they reported that the Jemez, the Keres (from Santa Ana and Zia), and the Tiwas (from Alameda, Puaray, and Sandia) had established themselves in the mountain strongholds of the Jemez (Hackett and Shelby 1942: 360). Vargas tried unsuccessfully for two years to bring the Pueblos down from their remote villages, but with the aid of Keres Pueblos loyal to the Spanish, Vargas ultimately captured and burned the pueblo after a violent battle in 1694 (Espinosa 1942: 200). With the exception of Astialakwa, Patokwa, Boletsakwa, and Cerro Colorado, the precise locations of the Jemez strongholds was unrecorded by Vargas.

The next section reviews the archaeological evidence supporting the continuous or intermittent occupation of villages in the Jemez Mountains. The following sections explain the natural environment and history of archaeological research in the Jemez region. I then present summaries and site descriptions of those pueblos known to have been occupied during the late sixteenth-century and Revolt periods. Finally, I compare this information with historic population estimates from the Jemez missions and argue that periodic reoccupation of Jemez sites should be factored into explanations of historic settlement and population shifts.

Environmental Setting: Drainage Systems and Geological Formations[2]

The Jemez Range and Plateau regions represent the southern extensions of the Rocky Mountain range. The two predominant geographic features of the area are the Valle Caldera, located midway between the Chama and Jemez rivers and the Rio Grande Rift Valley. The communities of the Upper Rio Grande Valley analyzed here are located on the southern edge of the Jemez or Pajarito Plateau and on the western side of the Rio Grande in the Jemez Mountains (Hewett 1906: 9), which designated the parameters of the Jemez Plateau as extending from Santa Fe, directly north and west to the contemporary Colorado border. The western edge is marked by the Rio Puerco, forming the western boundary of contemporary Jemez Reservation lands. The Rio Grande Valley forms the eastern boundary of the plateau.

The entire region is divided by three principal watershed and tributary systems. The Rio Chama runs northwest to southeast, draining the northern portion of the Jemez Mountains. The Jemez River drains the southern Jemez range, and the Northern Rio Grande Valley channels runoff from the Pajarito Plateau to the Rio Grande. These rivers form the three principal tributary watershed systems of the region and provide stable, year-round access to water in the semiarid northern Rio Grande region. The western Jemez range is framed by the Puerco to the west and the Jemez River to the east. The eastern Jemez range is similarly bounded by the Jemez and Rio Grande rivers.

The Jemez River is fed by a system of four main tributaries, each of which are located on the floors of steep and rugged canyon lands. The East Fork, San Diego Creek (or Rio Jemez), Rio Guadalupe, and

Vallecitos (Viejo) Creek form the Rio Jemez that joins the Rio Grande just south of contemporary Santa Ana and San Felipe pueblos.

Massive eruptions during the Pleistocene created extensive, multilayered rhyolitic ash deposits (Goff 1990). The simultaneous erosion and uplift of these geologic features help create the often dramatic elevation changes between the high, flat mesas and plateaus and narrow, steep-sided river valleys. The Jemez and Pajarito plateaus consist of a series of large tables and fingerlike mesa projections, many of which served as the sites of Revolt period communities. Steep canyons bisect these mesas with smaller, irregularly spaced fertile alluvial valleys. The difficulties involved in locating (much less ascending) the well-hidden, often perilous trails to the summits of these mesas attest to the difficult technical problems that confronted Puebloan communities during the Revolt and contact periods. While some mesa-top cisterns were available, both water and produce had to be brought up from the valley floors as many of the mesas lack significant water sources or cultivatable field areas.

Mountain range elevations reach 11,200 feet, and mesa elevations range from 6,500 to 8,000 feet. Higher mountain regions are heavily wooded with pine, spruce, and juniper, characterized by Elmore as the Fir-Aspen Belt (1976: 157). Mesa and valley regions are dominated by classic pinyon and juniper (P&J) forests. Both P&J and pine-oak (sometimes collectively known as Upper Sonoran) belts form an intermediate vegetation zone that gradually gives way to the more sparsely vegetated lower regions of the Rio Grande Rift Valley.

The presence of permanent water sources fed by runoff from the Rocky Mountains coupled with fertile volcanic soils creates conditions favorable to the development of a diverse yet tenuous subsistence base for Puebloan communities. Annual precipitation varies from 300 to 400 millimeters per year, with an average growing season (measured between first and last frosts) of 120 to 160 days per year (Elliot 1982: 2). The predominantly arid climate alternates with both winter and summer precipitation cycles. Winter storms generally develop in the Pacific to the west, while the classic summer monsoon season brings the majority of annual precipitation north from tropical systems originating in the Gulf of Mexico. Precipitation patterns during the summer months are often highly variable from season to season, and irregular, regional fluctuations are also quite common. Generally precipitation increases with elevation; annual readings from the southern Rockies average forty inches, while those of the Albuquerque Basin (ten inches) are considerably lower.

High standard deviations for both regions make dry farming a risky subsistence strategy for Puebloan communities. Regional variations are thought to have contributed to the spread of regional alliances and reciprocal relations between kin, clan, and separate Pueblo communities. Domesticates common to the region such as *Zea* maize require a minimum of twelve inches of annual rainfall as well as a frost-free growing season of 110 to 130 days (Rautman 1993: 403–424). Extreme diurnal temperature fluctuations are common throughout the year; summer temperatures can exceed 100°F, while subzero temperatures are common at higher elevations during the winter months. Subsistence strategies for Pueblo communities are well documented (Berry 1985; Cordell 1984; Dean et al. 1985; Euler et al. 1979; Glassow 1980; Minnis 1985; Simmons 1986)[3] and include the use of a combination of domesticated and cultivated crops, supplemented with a variety of wild game.

Faunal and Floral Resources

Fauna in the region included mule deer *(Odecoileus hemonius)*, bighorn sheep *(Ovis Canidensis)*, elk *(Cervis canadensis)*, bobcat bear, and coyotes, as well as smaller mammals such as rabbits *(Lepus californicus, Sylvilagus audubonii)*, squirrels, and prairie dogs *(Cynomys gunnisoni)*. Fish such as cutthroat trout *(Salmo clarkii)* and migratory and nonmigratory birds such as turkey *(Melcagris gallapavo)* and grouse *(Dendragapus obscuras)* were potential sources of food. Larger mammals such as antelope *(Antilocapra americana)* and bison *(Bison bison)* inhabit grasslands to the east of the Rio Grande Valley and were an important resource for Puebloan people as well as their neighbors (Trierweiler 1990).

Most importantly, high environmental variability coupled with relatively dense, permanent communities required careful, long-term subsistence strategies. Still, Puebloan people successfully established long-term occupations in an essentially marginal environment through the development of regional, reciprocal trade networks and carefully maintained, long-term dry storage systems located in the dry, cool, and darkened inner rooms of Pueblo settlements. Understanding the differences between prehistoric, regional *reciprocal* trade networks and the largely extractive, exploitative tribute demands of the Spanish colonists is absolutely essential in the examination of postcontact Puebloan political economy. As demonstrated in the previous chapter, food stores conserved and stored in anticipation of dry years were frequently targeted by Spanish colonists. In this sense, the presence of agricultural surpluses was often misinterpreted by Spanish colonists as a signal of seasonal abundance rather than an insurance policy against the inevitable dry and lean years that regularly occur in the region (Cordell 1984: 120).

While the droughts of the sixteenth century were severe, legally sanctioned and militarily enforced social hierarchies between colonists and Native peoples forced often unrealistic and unsustainable tribute and labor demands on Pueblo communities. At the same time, Pueblo people were forced to abandon carefully planned subsistence strategies that favored the use of widely scattered fields and hunting grounds. The travel of Pueblo people was highly restricted, and the policies of congregación and reducción forced the pueblos into more easily monitored, larger communities while freeing up large tracts of irrigable lands for the use of Spanish colonists. While these social factors are often overlooked in studies of population decline, they undoubtedly compromised the health of many communities—facilitating and accelerating the impact of diseases, crop failures, and famine on Indigenous populations.

Archaeological Resources in the Upper Rio Grande: Mission Studies as a Proxy for Puebloan Communities

Because of the close association between North American archaeology and the study of Puebloan prehistory, the establishment and refinement of upper Rio Grande chronologies has long been an important focus of archaeological research. As this study is concerned with contact and postcontact period research, I offer a simplified overview of regional shifts during the prehistoric period.

The focus of historic period archaeological research in the region has traditionally emphasized historic mission communities. Kidder's work at Pecos (1916, 1924, 1927, 1958) established a general regional chronology for Puebloan prehistory (see table 6.1). Studies of other mission settlements such as Hawikuh (Hodge 1918; Smith, Woodbury, and Woodbury 1966), Gran Quivira (Hayes 1981; Hayes, Young, and Warren 1981; Vivian 1964), Quarai (Hurt 1990), Abó (Toulouse 1949), and Awatovi (Brew 1937; Montgomery, Smith, and Brew 1949; Lawrence 1951; Smith 1952, 1971; Olsen 1978) have been particularly important in reconstructing the relationships between Spanish colonists and Puebloan peoples.

TABLE 6.1. Upper Rio Grande Glazeware Chronology

Glaze Group	Regional Ceramic Interpretations and dates				
	Mera (1940)	*Snow and Warren (1976)*	*Warren (1979)*	*Sundt (1987)*	*Schaafsma (1995;1996)*
Glaze A					
Los Padillas Glaze Polychrome	n/a	–1350	1300–1325	1300–1350	1315–1350
Arenal Glaze Polychrome	n/a	–1350	1315–1350	1315–1350	1315–1350
Agua Fria Glaze on Red	1350–1425	1325–1425	1315–1425	1340–1425	1340–1425
San Clemente Glaze Polychrome	1350–1425	1350–1400	1325–1425	1350–1400	1340–1425
Cieneguilla Glaze on Yellow	1350–1425	1350–1425	1325–1425	1375–1425	1370–1425
Cieneguilla Glaze Polychrome	1350–1425	1375–1425	1325–1425	1400–1450	1370–1425
Glaze B					
Largo Glaze on Red	1425–1450	1400–1425	n/a	1400–1425	1415–1450
Largo Glaze on Yellow	1425–1450	1400–1425	1400–1450	1400–1425	1420–1450
Largo Glaze Polychrome	1425–1450	1400–1425	1400–1450	1400–1425	1425–1450
Glaze C					
Espinoso Glaze Polychrome	1450–1490	1450–1475	1425–1500	1450–1475	1450–1500
Glaze D					
San Lazaro Glaze Polychrome	1490–1515	1475–1515	1490–1515	1475–1515	1490–1525
Glaze E					
Puaray Glaze Polychrome	1515–1650	1515–1625	1515–1600	1515–1625	1515–1625
Pecos Glaze Polychrome	1515–1650	n/a	1600–1700	1600–1700	1525–1625
Glaze F					
Kotyiti Glaze on Yellow	1650–1700	n/a	1650–1750	1600–1700	1625–1700
Kotyiti Glaze on Red	1650–1700	n/a	1650–1750	1600–1700	1625–1700
Kotyiti Glaze Polychrome	1650–1700	–1700	1650–1750	1600–1700	1625–1700

Very little other work has been done at other historic period Pueblo communities. Most historic period sites occupied by Puebloan peoples are either still occupied or are located on reservation lands. With some exceptions (Preucel 2000; Ferguson and Mills 1982; Ferguson 1996), collaborations between contemporary Pueblos and archaeologists have not been attempted until fairly recently. Many existing cultural resource management studies, excavations, or surveys of reservation lands have yet to be integrated into academic literature.[4]

Databases such as the Museum of New Mexico's Laboratory of Anthropology are the best source of information regarding regional settlement, survey, and site maps. Unfortunately, there are frequently considerable discrepancies between the more recent inventories, maps, and site data at the local agencies and the information housed at the central clearinghouse for archaeological research at the Lab of Anthropology. To remedy this situation, much of the archaeological information contained and organized in this book was collected from a combination of individual agencies (e.g., the Santa Fe National Forest offices) as well as the Lab of Anthropology files. Numerous site reports, published and unpublished in field reports, journals, and ethnographic materials published early in the twentieth century provide a substantial, if daunting, organizational task.

In addition to the archaeological data, the bulk of primary historical documents dealing with the early colonial and Revolt period history was published after much of the archeological work had already been done.[5] With the exception of John Kessell's reanalysis of life at Pecos Pueblo in the sixteenth and seventeenth centuries in *Kiva Cross and Crown* (1979), collaborative or interdisciplinary studies of the period have not been attempted. With only fragments of colonial historical documents at their disposal, the "first generation" of archaeological studies of Puebloan prehistory were often fragmentary—none seem particularly concerned with representing the plight of the Pueblos from a perspective other than that of military conquest and eventual defeat.

As a result of this information gap, archaeological studies of a only a handful of New Mexico's mission communities have generally served as a proxy for the evaluation of historic period Puebloan behaviors, economic activities, mortuary analysis, and disease profiles as well as measures of acculturation. In emphasizing the study of communities most closely supervised by Franciscan missionaries and ignoring those associated with movements of resistance, historic period scholarship tends to emphasize a more universally interdependent vision of social interaction and mutual accommodation than likely existed in the colonial period. In addition, as demonstrated by the sequences outlined here, little attention has been devoted to postcontact changes in the region.

Archaeological Approaches to Precontact Settlement Shifts in the Pueblo Region

Because the archaeological manifestation of the Revolt consists of both the abandonment of sites as well as communal aggregation and resettlement, the documentation and analysis of a known abandonment as well as the aggregative event that follows are of particular interest to archaeologists. Explanations for the causes of aggregation during the late classic and historic periods are numerous, but most attribute the process to a combination of economic, evolutionary, environmental, or conflict-driven causes (Adler 1994: 85–101, Adler et al. 1996a: 375–438; Crown 1994: 109). The need to establish cooperative alliances to ensure survival in a marginal landscape is frequently cited as the driving force behind aggregative population shifts (Dean 1985; Hill and Trierweiler 1986; Plog et al. 1988). The presence of similar architectural features, plazas, and ritual objects seems to support the notion that aggregation was accompanied by the spread and increased participation in a general suite of religious practices that characterize what has come to be known as Pueblo Religion (Crown 1996: 201; Cordell 1996: 237).

The presence of agricultural features associated with more intensive and productive subsistence strategies generally increases during this period. Although they are nearly impossible to date, canals, water diversion devices for rainfall accumulation in remote areas, and grid and cobble gardens have been observed at sites occupied during this period (Cordell 1984: 205, Cordell et al. 1994; Lightfoot 1993; Lightfoot and Eddy 1994, 1995; Woosley 1980).

Adams (1991) has similarly tied the aggregative processes of the late classic period to the development and spread of the Pueblo katchina cult. Ceramic vessels, murals, and rock art shrines loosely associated with the use of katchinas generally have been assumed to have appeared at this time as well. Other archaeologists have suggested that the diversity of localized pottery styles is reduced during this period, with a limited suite of patterns, motifs, and decorative technologies (Cordell 1984: 337; Crown 1996: Habicht-Mauche 1993, 1995). The analysis of source materials at late classic period sites seems to support a scenario in which the Pueblo world experienced increased interaction manifested in the exchange of decorative vessels.

Other scholars have suggested that the arrival of Athapaskan speakers from the north occurred in the decades preceding Spanish contact with the Pueblos (D. Gunnerson 1956: 346–365; J. Gunnerson 1979: 162–184). The significance of this event, whenever it occurred, is not entirely clear. The assumption among many prehistorians seems to be that the arrival of Athapaskans signaled a significant shift in the economic, political, and social world of the Pueblos.

Although there appears to have been intermittent warfare between some parties of Athapaskans (variously referred to as Teyas, Querecheros, or Navajos) and individual pueblos (Haas and Creamer 1996: 205–213; LeBlanc 1999), Pueblo communities had already experienced both conflict (LeBlanc 1999) and cooperation and exchange with communities within their immediate world; and they had well-established systems of communication, information exchange, and differentiation both within and outside what has come to be known as the Pueblo world. To assume that the Pueblos were isolated from other, more numerous and powerful neighbors prior to the arrival of Europeans or Athapaskans is an often implicit, if not poorly founded, assumption among contact period archaeologists. Contact with and knowledge of Plains tribes to the east, traders from the Aztec Empire, and communities in northern Mexico, Shoshonean-speaking peoples, O'Odham, and, for people at Zuni, Yuman-speaking peoples of the Upper and Lower Colorado are well documented (Habicht-Mauche 1987, 1988; Lange 1979; Lintz 1991; Riley 1987, Spielmann et al. 1990). Still, the need for defensible communities is frequently cited as an explanation for aggregation (Haas and Creamer 1996).[6]

Based on a variety of archaeological evidence, it seems likely that the Pueblos experienced relatively balanced (although not necessarily amicable) relations between various Pueblo and non-Puebloan communities prior to contact with Europeans (Spielmann 1983, 1991a, 1991b, 1996). In general, the establishment of larger aggregated settlements located near permanent water courses, extensive periods of occupation, increases in katchina symbolism as revealed in artwork, as well as ceramic homogeneity support the notion that the Pueblo region had achieved some kind of balance within and between communities. There is little evidence to support the ascendancy of any particular community in relation to the collective of other communities. While some, such as Pecos, appear to have served as regional trade centers or buffers between the Pueblos and their neighbors (Spielmann et al. 1990: 745–765), one can assume that linguistically affiliated communities formed the basis for broadly defined ethnic communities. Overall, the frequency and number of all types of *recorded* Puebloan sites decreases during the historic period. Many of the large classic period sites have continued to be occupied since the arrival of the Spanish, although with the colonial missionary policies of reducción and congregación, the mobility and freedom of movement of those Pueblos located near Spanish settlements were drastically reduced.

Early Archaeological Studies in the Jemez Mountains

Much of the known information regarding the settlements of the Jemez Mountains was collected in the early twentieth century by a combination of private, amateur collectors and researchers at the School of American Research and the University of New Mexico (UNM).[7] Beginning in 1928, UNM established a field school at Battleship Rock and used this as a base from which larger excavations of Unshagi (between 1928 and 1935) and Nanishagi (in 1938) were carried out. Reiter's (1938) documentation of earlier ethnographic and ethnohistorical work produced by Bandelier (1892), F. W. Hodge (1910, 1916), J. P. Harrington (1916), and Lansing Bloom (1922, 1931, 1938) is still the best to date. In addition, the collaborative work of dendrochronologist W. S. Stallings; anthropologist C. Kluckhohn; archaeologists E. L. Hewett, Anna O. Shepard, and G. Vivian; and historians such as France Scholes and George Kubler helped document over sixty sites in the region (1938: 20–21). Still, at the time, few if any of the Spanish documents contained in the Coronado series (with the exception of Ayer's [1916] Benavides memorial) had been published. Unfortunately, the ambitious, interdisciplinary work begun by Reiter was largely abandoned by the 1940s.

With the exception of Elliott's Forest Service reports (1982, 1993), more recent synthetic analyses of contact and historic period settlement patterns have not been attempted due to a variety of methodological, logistical, and theoretical factors. Difficulties involved in establishing adequate ceramic chronologies for Jemez Black-on-White wares (discussed at length later) have made regional analyses of large sites a complicated task. The relationships between these communities and the almost ubiquitous, smaller field house–like structures has long puzzled archaeologists working in the region. The presence of these smaller, independent, household-size structures has complicated attempts to estimate population figures in the area. Leibman's work (2005) represents an important new trend in the archaeological work of the area.

Archaeologist Robert Preucel (2000a, 2000b, 2005, 2006) can be credited for initiating much of this groundbreaking wave of research. Preucel (at Cochiti) along with former Zuni tribal archaeologists T. J. Ferguson (1996, 2002), Barbara Mills (2002), Kent Lightfoot (2004), and Chip Colwell-Chanthaphonh (2007) have been at the forefront of a recent trend in engaging Native peoples in archaeological research. Collaboration is of the utmost importance in gaining access to and doing research in any of these historic sites. Still, the process of collecting data has been slow and difficult. Logistical problems abound in supporting large-scale excavations in many of these isolated sites. Ironically, NAGPRA has complicated the process of conducting excavations in clearly affiliated ancestral Pueblo communities. Consultation and negotiation with tribes on nonprivate lands can be a lengthy process. Given the lack of communication between archaeologists and Pueblo communities in the past, many Pueblo communities are reluctant to allow archaeological work of any kind in ancestral villages—most, if not all, of which contains shrines still used by contemporary Pueblos. The combination of these factors has generally hindered long-term projects and excavations in the region since the 1980s.

Still, the UNM and School of American Research monographs of the 1930s and 1940s (Alexander and Reiter 1935) helped establish much of what archaeologists know about the prehistory of the region. Paleo-Indian sites are difficult to distinguish, and according to Elliot (1980), none have yet been identified in the Jemez Mountains. More clearly identified occupation begins in the archaic period, and Jemez Cave (another UNM field school site) contains some of the earliest-known (2440 ± 250 BC) samples of *Zea mays* in the Southwest (Alexander and Reiter 1935; Ford 1975: 22). Little is known about habitations prior to the coalition period, but as Elliot has emphasized, occupations prior to this date are likely masked by the proliferation of classic period sites (1982: 2), and surveys above eight thousand feet have never been attempted. Following the regional pattern, classic period sites are located near permanent water sources or on the tops of mesas. Associated with these medium- to large-sized pueblos are the numerous field houses found scattered intermittently throughout the region. Many of these larger, classic period settlements are located in the most inaccessible and isolated areas of the northern Rio Grande.

As demonstrated later, many of these settlements were reoccupied during times of conflict with Spanish officials in an effort to maintain social and geographic distance between the groups. Movement to these areas was a strategy deployed by the Pueblos and widely adopted during the Revolt in an effort to conceal religious practices from missionaries, protect material and human resources, and defend against military and social violence directed toward Puebloan people.

Historic Period Ceramics: Chronological Frameworks of the Rio Grande Glazewares, Jemez and Tewa Series

Because of the diversity of community representation found at Revolt and, to a lesser extent, historic period Pueblo communities, several overlapping chronological frameworks must be integrated and examined to understand the postcontact population shifts and community settlements. The following discussions of glazewares and Jemez and Tewa series ceramics will help clarify my discussion of Revolt period settlement patterns and population movements. Late classic and historic period ceramic typologies of the Upper Rio Grande have generally been associated with glazewares. Early type collections were assembled by Nelson (1916), Kidder and Shepard (1936), and Mera (1933). These early attempts to refine spatial and temporal classification systems were largely descriptive in nature and used bowl rim shapes to establish a very general relative dating sequence.

Later, H. P. Mera, a vocational archaeologist, produced the most extensive and enduring regional synthesis of Upper Rio Grande glazewares. Despite the fact that significant variations exist in temper materials (Kidder and Shepard 1936; Warren 1979; Cordell 1991), to date no regional examinations of source materials have been attempted. Considerable overlap of glaze types has been observed as several sites (Cordell and Earls 1984; Hayes 1981), and Creamer and Renken's (1994) reanalysis of excavated materials reveals considerable overlap of early glazes (A, B, C, D, and E) once they have appeared in a sequence.

A general sequence of glazewares has been correlated with tree ring dates at several sites, and the general chronologies developed by Warren and Snow (1976) and Warren (1979), have been summarized in table 6.1 and simplified in table 6.2. Late Glazes E and F, often associated with Spanish artifacts (Nelson 1914, 1916; Lambert 1954, 1981; Toulouse and Stephanson 1960), are clearly historic and are

TABLE 6.2. Simplified Glazeware Chronology

Mera's Period (1940:5)	Glaze Types	Likely Dates
Period 1	Groups A, B	1350–1450
Period 2	Group C	1450–1490
Period 3	Group D	1490–1515
Period 4	Group E	1515–1650
Period 5	Group F	1650–1700

NOTE: After Schaafsma 1995.

the chief source of temporal identification among the Revolt period sites discussed here.

Wide acceptance of the glazeware index is partially the result of the lack of information provided by Jemez Black-on-White ceramics. The Jemez materials are notoriously difficult to seriate based on temper, shape, or design motifs (Crown, Orcutt, and Kohler 1996: 191). Because of the difficulties associated with Jemez wares as well as the logistical challenges imposed by the isolation of many of these isolated communities, few excavations have taken place at many of these more remote postcontact period sites.

In addition to the glazeware chronologies, Tewa ceramic types are also useful indices of temporal variability (see tables 6.3 and 6.4). Santa Fe and Kwahe'e Black-on-White indicate late developmental and early coalitional (1025–1300 AD) occupations. Wiyo or Santa Fe Black-on-White (Biscuitoid) and Abiquiu Black-on-Gray (Biscuit A)

TABLE 6.3. Tewa Ceramic Series

Upper Rio Grande Pottery Types	Dates
Jemez Black-on-White	1350–1700
Tewa Polychrome	1650–1730
Sankawi Black-on-Cream	1520–1600
Potsuwi'i Incised	1450–1600
Bandelier Black-on-Gray "Biscuit B"	1400–1550
Abiquiu Black-on-Gray "Biscuit A"	1360–1450
Wiyo Black-on-White "Biscuitoid"	1300–1400
Wiyo/Santa Fe Black-on-White "Transitional"	1275–1325
Santa Fe Black-on-White	1175–1300
Kwahe'e Black-on-White	1025–1175

TABLE 6.4. Tewa Ceramic Series Descriptions

Tewa Series Types	Dates	Notes
Tewa Polychrome	1650–1730	Tewa Polychrome and all of the post-1600 types were lacking on the Chama Valley Tewa sites according to Mera (1934) and were lacking at Te'ewi (Wendorf 1953: 54–57). Tewa Polychrome was lacking at San Gabriel, occupied to 1610 (Ellis 1989). Harlow (1974) offered the beginning date of 1650 (1974).
Sankawi Black-on-Cream	1520–1600	Beginning date is partly based on the presence of this type at Te'ewi in fill of Room 13 with a 1529 tree ring date (Dean, personal communication, 1996; Wendorf 1953:59), and indications that Te'ewi was abandoned by 1540.Wendorf noted that all Sankawi sherds at Te'ewi were of the early variety (Wendorf 1953:55). Since all of the survey data refer to this kind of pottery (e.g., Hera 1934), this definition is retained even though several efforts have been made to refine the type.
Potsuwi'i Incised 1953:55–56).	1450–1600	Wendorf and Mera thought it began in early Biscuit B times (Wendorf
		Sundt agrees with 1450 beginning date (1987:139); it may have still been made after 1598 at Yunque-Yunque, hence the 1600 termination estimate.
Bandelier Black-on-Gray "Biscuit B"	1400–1550	Construction date of of Kiva I at Te'ewi was near 1411 based on tree ring samples on the floor (Wendorf 1953:93). The presence of Biscuit B in the subfloor tests means this type was in existence by then.
Abiquiu Black-on-Gray "Biscuit A"	1360–1450	Beginning date is based on Las hadres Stratigraphic position. (Schaafsma 1995); end is based on Sundt (1987:136).
Wiyo Black-on-White "Biscuitoid"	1300–1400	Early, well-developed Wiyo is firmly dated at the Palisade Ruin.
Wiyo/Santa Fe Black-on-White	1275–1325	This is an approximation of a long period of transition.
Santa Fe Black-on-White	1175–1300	Definite Santa Fe Black-on-White, not the transitional kind. Wiyo was probably not made after about 1300 (Sundt 1987:140).
Kwahe'e Black-on-White	1025–1175	Very little Kwahe'e in the Chama Valley. Refer to Sundt (1987:138) for the reference dates.

are indicative of late coalition, early classic (1275–1450) occupations (Schaafsma 1995; Sundt 1987: 136). According to tree ring samples collected from Wendorf's excavations of kiva 1 at Te'ewi (1953: 93), Bandelier Black-on-Gray (Biscuit B) mark the beginning of the classic period (1400 AD) and terminate around the time of Coronado's entrada in 1540.

The later Tewa series ceramics are particularly valuable in determining the relative dates of Revolt and contact period occupations. Potsuwi'i incised ceramics are found throughout the classic period (1540–1600) and are an excellent indicator of entrada period settlements. Likewise, Sankawi Black-on-Cream (1520–1600), identified by Wendorf and dated using tree ring samples from Te'ewi (1953: 55), can be considered accurate temporal markers of entrada period settlements. Of more immediate importance are the classic markers of Revolt period occupations, the Tewa polychromes. According to Ellis (1989), Tewa polychrome materials at are absent from San Gabriel (San Gabriel is also known as Yungue, Yungue-Yunquee, and Yuge-uin-ge).[8] San Gabriel was partially abandoned after Oñate forced out Tewa inhabitants and established the first provincial capital of New Mexico in the center of the Tewa Village complex. The appearance of Tewa polychromes roughly coincides with the establishment of a permanent colonial presence in New Mexico, although the end dates for this series (1730) are not particularly helpful in determining the length of occupation at Revolt period sites.

Historic Period Villages in the Jemez Region: Site Descriptions and Analysis

Jemez region sites can be categorized in several different ways. Table 6.5 and figure 17 present those sites that have been identified by Forest Service archaeologist Mike Elliott (1982) as being located within the Jemez province. These large settlements can be split into two groups. The first four—Cerro Colorado, Patokwa, Astialakwa, and Boletsakwa—are large villages that were built during the late 1600s. As demonstrated by the pottery types present at the site, these communities may have been the sites of earlier, ancestral villages constructed during the mid- to late 1600s. However, the later villages appear to have been constructed and occupied during the years of the Revolt of 1680. Limited samples of tree ring dates analyzed by Robinson et al. (1972) have provided some sense of a possible construction sequence, although these samples only reinforce the dates provided by ceramic samples.

TABLE 6.5. Historic and Revolt Period Sites in the Jemez Region

LA Number	Site Names	Estimated Rooms	Glazewares	Ceramic Dates
2048	Cerro Colorado	150	E-F	1515–1625/1700
96	Patokwa, San Diego del Monte	600	E-F	1515–1625/1700
1825	Astialakwa	250	E-F	1515–1625/1700
136	Boletsakwa	650	E-F	1515–1700
303	Seshukwa, San Juan Mesa Ruin	1100	D-E-F	1490–1700
482	Kwastiukwa, Giant Footprint Ruin	2750	D-E-F	1490–1700
LA 132, 133	Kiatsukwa (?)	325	D-E-F	1490–1700
481	Amoxiumqua, Virgin Mesa Ruin	1200	A-B-C-D-E-F	1350–1700
484	Tovakwa, Stable Mesa Ruin	1850	D-E-F	1490–1700
2049	Tamaya	30	E-F	1515–1700
135	Little Boletsakwa	350	D-E-(F?)	1490–1700
483		300	C-D-E	1450–1625/1700
398		450	D-E	1490–1625/1700
44000		250	D-E	1490–1625/1700
44001		125	D-E	1490–1625/1700
46340	Kiashita	50	D-E	1490–1625/1700
24788	Wahajhamka	750	D-E	1490–1625/1700
5918		525	D-E	1490–1625/1700
24790		125	E	1515–1625/1700
541	Nanishagi	450	D-E	1490–1625/1700
123	Unshagi	300	D-E	1490–1625/1700

(Continued)

TABLE 6.5. *(Continued)*

LA Number	Site names	Estimated Rooms	Glazewares	Ceramic Dates
479	Totaskwinu	300	D-E	1490–1625/1700
5920		475		
478	Wabakwa	1400	A	1300–1425
24553	Hot Springs Pueblo	75		
385, 5928		75		
386		150	A	1350–1425
130	Pejunkwa (?)	1300	A	1350–1425
189	Guacamayo Ruin, Kiabakwa	400	A	1350–1425
24689		75		
24792	Hanakwa	75		
137		100		
128		125		
403		75	A	1300–1425

FIGURE 17. Historic and Revolt Period Villages in the Upper Rio Grande.

The other category of sites—Seshukwa, Kwastiukwa, Kiatsukwa, Amoxiumqua, Tovakwa, and Tamaya—differentiated by the presence of earlier glazewares (A–F), are those that were reoccupied throughout the seventeenth century. The presence of glazewares at these villages supports the notion that ancestral Jemez settlements were periodically resettled during periods of conflict. This supports two arguments presented in this contextual study. First, the policies of reducción and congregación as well as the attempts by Franciscan missionaries to contain the activities and movements of the Jemez and their allies were only marginally successful. Large ancestral sites, far from the missions of the Jemez River Valley, housed and supported Jemez people (and presumably their allies) who were unwilling to accept the forcible conversion attempted by the missionaries. The evidence supporting the reoccupation of these villages raises important questions: How long were these settlements occupied? How many of the Jemez people were involved in these movements? How many other large sites in the Pueblo region demonstrate similar reoccupations? Finally, how might similar resettlements challenge the population crash models so frequently deployed in historic period studies? Clearly, a more nuanced understanding of population shifts is required to understand how mobility was used as a strategy of boundary maintenance and resistance.

Two distinct strategies of resistance are evidenced by these villages. Both categories of settlements support the notion that both aggregation and dispersal were tactics adopted by the Jemez to escape surveillance by both colonists and missionaries. At Cerro Colorado, Patokwa, Astialakwa, and Boletsakwa, the Jemez aggregated themselves into large, defensible mesa-top villages. In contrast, the reoccupations at Seshukwa, Kwastiukwa, Kiatsukwa, Amoxiumqua, Tovakwa, and Tamaya provide evidence supporting the dispersal of the Jemez and their allies. The following section presents the collective information provided by Forest Service reports as well as earlier ethnographic and historical work. Much of the archaeological data and fieldwork was collected by Mike Elliott in the early 1980s (1982) as well as several earlier archaeological and Forest Service and Museum of New Mexico reports. The final section of the chapter compares this information with historical population figures. This comparison suggests that many of the historical estimates do not support a disease-based population crash model and should be used with caution.

Jemez Region Historic Period Site Information

CERRO COLORADO "RED MESA," "MESA COLORADO" (LA 2048)

Previous Research. Reagan 1922; Hewett 1906: 46; Holmes 1905: 201.

Location. Several miles west of Jemez Pueblo on an isolated mesa west of the Rio Salado.

Site Description. Cerro Colorado consists of three distinct roomblocks (labeled A–C on figure 18). Roomblock A, the largest of the three, is a sprawling C-shaped arrangement of ninety-five to one hundred connected rooms enclosing two central plazas. The northern segment of the structure curves along the northern edge of the mesa in a semicircular cluster of twenty-five to thirty ground floor–level rooms. The western portion of roomblock A contains twenty-five to thirty ground floor rooms and stretches to the south in a linear, north-south orientation. To the west is a spur of ten additional rooms. These inner rooms

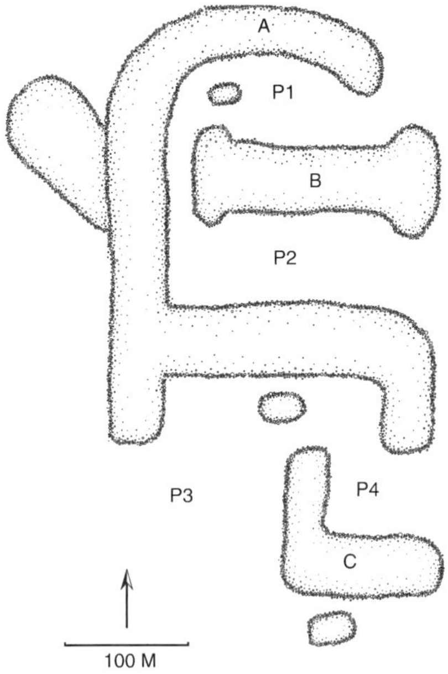

FIGURE 18. Cerro Colorado.

are isolated from the plazas and may have supported additional stories. The southern portion of roomblock A contains thirty- to thirty-five ground-level rooms. Two southern extensions of this roomblock frame plazas 3 and 4. An isolated room or possibly a shrine is located a few feet south of this segment of the structure in plaza 3.

Roomblock B bisects the open area between the northern and southern portions of roomblock A, forming two central plazas. This barbell-shaped roomblock is separated from roomblock A by a narrow, western passage. The western segment of roomblock B contains a cluster of inner rooms and may possibly have held a second story. Just north of this roomblock in plaza 1 is a separate structure that may have been an isolated room or possibly a shrine. Ground floor–level rooms are estimated at twenty-five to thirty.

Roomblock C is an L- shaped structure located twenty to twenty-five meters south of roomblock A and contains thirty-five to forty rooms. The north-south alignment of rooms separates plazas 3 and 4, and the east-west line of rooms closes off plaza 4 from the southern cliff face. A small, single-room structure or shrine is positioned just south of the east-west segment of roomblock C. The village stretches approximately 150 to 170 meters from the northernmost portion of roomblock A to the southernmost portion of roomblock C. The village stretches approximately 125 to 150 meters, east to west.

The walls are constructed from shaped tuff and mortar and stand from 1 to 1.5 meters in some places, although Holmes characterizes the masonry as consisting of "rough, loosely laid stones" (1905: 200). Despite the apparent size of the village, the walls are noticeably thin and unstable in places, although wall alignments are described as clearly discernible. There are no circular or "kiva-like" structures in any of the plazas or in the immediate vicinity of the site.[9] There are several defensive structures, small walls, and piles of stones at the exposed or more easily accessible portions of the mesa.

Room Number Estimates. 150–170 ground floor–level rooms are present at the site.

ARTIFACTS AND CERAMIC DESCRIPTIONS

Ceramics present include Agua Fria Glaze on Red, Espinosa Polychrome, Puaray Glaze on Red, and Jemez Black-on-White, Biscuit B (Mera's notes mention unspecified Hopi sherds as well). Glazewares E and F predominate. Obsidian flakes and cores as well as vesicular basaltic manos and metates are present also. Artifact density is noticeably low for a site of this size.

PATOKWA (LA 96, FS 5)

Previous Research. Bandelier 1892; Holmes 1905; Reiter 1938; Scholes 1938; Elliott 1982, 1989.

Location. Patokwa lies on a small bench at the confluence of two small rivers. The village is located near several cultivated fields; willow and riverine flora as well as P&J bushes are present.

Site Description. Patokwa is a fairly large (100 × 60 meters), rectangular pueblo (figure 19). Two main plaza areas are bisected by a single roomblock, separating both halves of the village. Two possible small kivas are located in the south-central portions of both plazas, and a

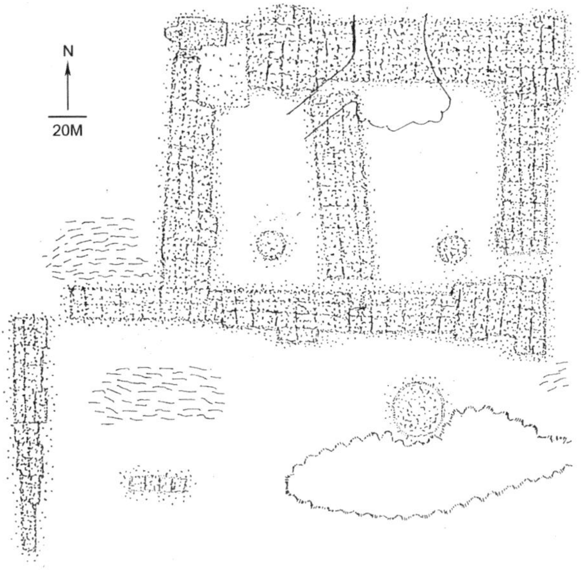

FIGURE 19. Patokwa.

larger kiva was apparently located just outside the southern roomblock. Two contiguous L-shaped roomblocks (one oriented north-east and the other south-west) frame both plazas, revealing openings in the northwest and southwest corners. There is evidence of a small church in the outer northwest corner of the pueblo, possibly identified as the San Diego del Monte Mission The mission was in use between 1694 and 1696 but was destroyed by the Jemez after the Revolt of 1696. Roomblocks were likely three to four deep throughout the main section of the village, although the southwest roomblock appears to have contained fewer rooms. A smaller roomblock is situated outside the southwest corner of the village. This north-south-oriented section of rooms stretches thirty to forty meters and is one to three rooms deep. This portion of the site frames a third possible plaza area that opens to the south. One smaller roomblock segment (five to ten rooms) is located south of the southern roomblock. A separate section of rooms lies just east of the eastern roomblock. The north-east, south-west, and bisecting roomblocks appear to have supported at least a second story. The northern roomblock has been heavily damaged by a bulldozer, and the site has been heavily damaged by illegal excavations.

Room Number Estimates. Wall heights and rubble sizes are consistent with a multiple-storied pueblo. Room numbers at the site have been estimated at 600 to 650.

Artifacts and Ceramic Descriptions. Artifact density is fairly high, although illegal collections appear to have been a serious problem at the site. Lithic samples reveal several obsidian and chert flakes. Basalt fragments and cores are present as well. The ceramic collections contain Jemez Black-on-White; plain, corrugated, and smudged utility wares; European wares; as well as Glazewares E and F.

ASTIALAKWA, "GRINDING STONE LOWERING PLACE" (LA 1825, FS 360)

Previous Research. Holmes 1905; Reiter 1937; Espinosa 1942; Dougherty 1980; Elliot 1982.

Location. Astialakwa lies at the southwestern edge of a large prominent mesa in the Jemez region. The mesa is capped with white volcanic tuff and gradually slopes from a northern prominence of 6,968 feet to the southern edge, where the village of Astialakwa is located. The village is situated nearly 250 feet lower than the summit and is positioned to take full advantage of runoff from the riparian streams on the mesa top.

The elevation changes between the summit mesa and the surrounding landscape are dramatic; a small river runs through a fertile floodplain at the base of the canyon is 920 feet below. The site is positioned about 300 meters from the southern edge of the mesa. At 6,680 feet, Astialakwa is accessible by a main trail in the western face of the mesa. The trail winds through several nearly vertical exposed rock outcroppings. Defensive features along the trail include a series of small stone fortifications invisible from the approach below. A series of smaller activity sites are scattered throughout the mesa, many of which are at higher elevations.

Two main riparian water sources are fed by the runoff at higher elevations on the mesa. The easternmost stream bisects the main village, creating two distinct habitation areas: FS 325 is to the west of the stream, and FS 317, FS 326, FS 323, and FS 319 are located to the east of the stream. Water diversion devices are present at several locations on the mesa.[10] One (FS 294) is located two hundred meters north of the main village (FS 325). The riparian stream bisecting the main village has several diversion and catchment modifications, and the main garden area (LA 280) contains two check dams that help divert water from a nearby stream. Caveate structures and cliff dwellings are scattered throughout the northeastern, eastern, and southwestern edges of the mesa, and several stone walls used as defensible observation points are present in the northeast-, southeast-, and southwest-facing cliff areas.

Arable lands are located near the summit where P&J stands predominate. A cleared area of grid gardens lies just south of the summit, and cultivatable fields in the valleys to the east and the west of the mesa are extensive. Shrines are also scattered throughout the mesa. Astialakwa and the Guadalupe Mesa continue to hold special significance to the Jemez people and are monitored continually by tribal and Forest Service personnel.

Site Description. Astialakwa consists of a series of irregularly arranged roomblocks that form roughly four general clusters: FS 325, 317, 326, and FS 323/FS 319 (see figure 20). FS 325 is the largest cluster of rooms and is located to the west of the arroyo. Consisting of approximately 170 to 180 ground-level rooms, the structures follow a pattern typical of some contemporary Pueblo communities; the roomblocks are arranged in six to seven parallel rows with depths of one to three rooms each. While roomblocks contain between two and thirty rooms, many of the structures are single- or double-room features, set off from the surrounding roomblocks.

Frequently, isolated single or double rooms were not architecturally integrated into the other roomblocks. This suggests that although Astialakwa was occupied for several years, the construction of the site seems to have been less organized or coordinated than the construction of other similarly sized pueblos.

Room sizes are highly variable and range from 2 × 3 to 4 × 6 meters. Construction materials consist of shaped volcanic tuff and unshaped sandstone blocks set in coursed rows. Mortar and small stones are used for chinking, and vents, doors, and plaster are in evidence at several rooms. Rooms toward the southwest of the village are set into the side of the cliff and utilize bedrock for wall and floor surfaces. Viga holes above and below these rooms suggest that multiple stories were present, and a stairway, carved out of the rock face, leads down to these rooms from the

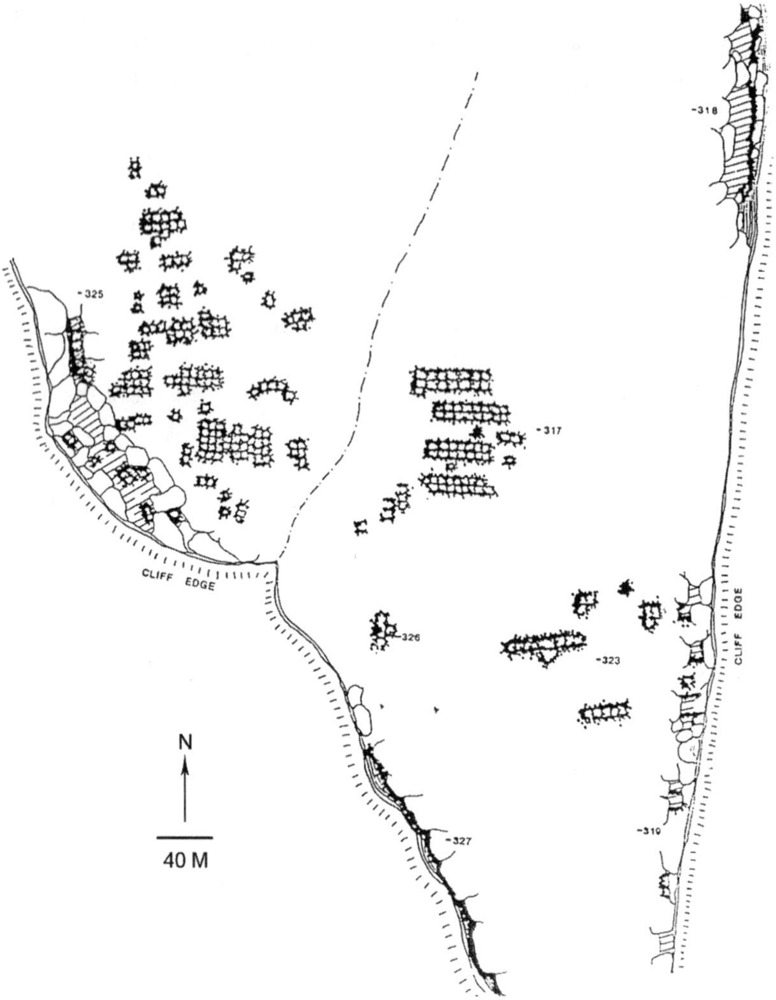

FIGURE 20. Astialakwa.

main habitation area. A series of very narrow passageways in the jumble of large stones at the mesa's edge link the cliff face roomblocks. While there is a semicircular arrangement to rooms just west of the arroyo, there appears to be no main plaza area on Astialakwa. There are also no circular rooms or mounds commonly identified as kivas.

FS 317 is the second habitation area on the south face of the canyon. Located forty meters to the east of FS 325 across the main arroyo, the site consists of four to five parallel rows of roomblocks, each of which is two to three rooms deep. Average room sizes are a bit smaller than FS 325 (2 × 3 meters). Building materials are identical to those used at FS 325, and there are no discernible differences in average stone size or the use of coursed masonry or plaster. There is a small cluster of rooms, set off from the main concentration of roomblocks just southwest of the parallel rows. These four rooms share no walls and are not architecturally linked to the other rooms. To the east of FS 317 are some burned areas, close to the edge of the mesa. These may have been used as signaling areas for other pueblos or to people in the valley below. Because of the prominent location of the village, visibility can extend forty to sixty miles on many days. Because of its southern-facing exposure and the fact that the Spaniards usually approached the Jemez area from the south, Astialakwa must have been an important defensive outpost or first line of defense for the other Revolt period Pueblos in the Jemez area.

FS 326 is a separate habitation area, sufficiently isolated from the other rooms to be considered a separate site. This five- to six-room structure is located forty meters south of FS 317. Room sizes are approximately 4 × 6 meters, and building materials are similar to those at the other habitation areas. Coursed masonry with mortar and small stones are built directly on top of the bedrock. There are no subterranean or circular kivas associated with this site.

FS 319 is located thirty meters southeast of FS 317 and twenty-five meters east of FS 326. The site is composed of four roomblocks arranged in parallel rows. The central roomblock contains nine to ten rooms, the southernmost roomblock in the cluster contains five to seven rooms, and the others are single or two- to three-room clusters. Building materials are consistent with those from the other habitation areas. A significant difference between the central, ten-roomblock habitation is the size of the rooms. Whereas many of the rooms throughout the site are 3 × 5 meters in size, the center two and far right rooms are significantly larger (4–5 × 6–7 meters) than the others. There is also an eight- to ten-roomblock dwelling built into the eastern-facing cliff edge.

Just south of these rooms is FS 319, a similarly shaped, smaller cliff dwelling built on a shallow bench at the edge of the mesa. The only other caveate structures recorded at the site is FS 296, a small dwelling located approximately 0.7 to 0.8 kilometer directly north of FS 325 on the northern edge of the mesa. Built directly into the cliff, a volcanic tuff, masonry doorway helps diminish exposure to wind and sun. There is also evidence that a kind of ramada-like structure helped obscure or shelter the opening to the rooms. Viga holes and small niches are built into the walls of these fire-blackened small rooms.

Defensive Fortifications. Nearly the entire northeast quadrant of the cliff face is bordered by a low-walled fortification (FS 279, 278, 324, 321, and 322). The southern edge of the mesa is highly fortified as well (FS 318, 319, 327, and 295). Walled fortifications are present at nearly all accessible positions on the approach to the mesa as well as in areas where the mesa slopes more gradually from the cliff edges. Defensive walls are generally built of uncoursed masonry and are constructed on areas of exposed bedrock, and some are perched on naturally occurring benches directly beneath the cliff edges. Stone sizes are similar to those used in the construction of walls in the habitation areas. Many of the lower stones are between 0.5 and 0.75 meter in length. Wall heights range from 1 to 2.5 meters. The lower wall heights are likely the result of natural erosion and site deterioration as well as vandalism. Much of the site was destroyed in the attacks by Spanish colonists in the years following the revolt of 1680.

At FS 322, there are circles of blackened stones at the edge of the mesa. This site is approximately 0.75 kilometer north of the "signal fire" area of FS 319 and FS 323. At the southwestern trailhead are stacks of stones piled behind the south facing wall (FS 327). A similar stack of stones, piled behind a wall at the northern approach to the mesa (FS 27), signals the existence of a trail on the northern face of the mesa, but this trail was not surveyed. At both defensive positions the stones were apparently selected for their weight. River cobbles, vesicular basalt, sandstone, and quartzite stones were collected and piled in these areas.

Water Control Features and Gardens. FS 280 is located near the northern summit of the mesa just to the east of FS 278 and FS 324. The site consists of a series of collapsed storage cists, a check dam, and a collection of grid gardens. Located south of a small stream and running through the center of the gardens, the two check dams are constructed of volcanic tuff cobbles and boulders. The grid gardens are situated in a cleared area of P&J forest. There are twenty to twenty-five 2 × 4-meter semirectangular gardens. Each is bordered with shaped volcanic tuff and are generally oriented east to west in parallel rows. All of the grid garden rows run east west and are positioned to collect the runoff from the garden above.

FS 294 is a water and soil control feature situated 250 meters north of the main habitation area of FS 325 and FS 317. This feature is a mounded berm of unshaped volcanic tuff and cobbles loosely piled into a low thirty- to forty-meter-long, semicircular wall. Soil collects on the upslope and downslope sides of the wall and may have been used as both a water storage and agricultural feature.

Room Number Estimates. The following should serve as a rough estimate of the number of rooms present at Astialakwa. It is important to remember that without excavations, the true number of rooms may never be known. Accurate assessment of the amount of rubble in rooms as well as the presence of hearths would help illuminate the number of double- or multiple-storied rooms. FS 325 contains 180+ ground-level rooms, FS 317 has 60 ground-level rooms, FS 326 has 5 ground-level rooms, and FS 323/319 has 35 rooms. The total estimated number of ground-level rooms is 280.

Artifacts and Ceramic Descriptions. Jemez Black-on-White, Glazes E to F, plain and corrugated utility wares, Salinas Red, and Tewa Polychromes are also present. Notes at the Lab of Anthropology describe numerous large olla fragments, suggesting that the storage of water was a significant concern. Lithics include obsidian, basalt, and chalcedony flakes. Basalt mano and metate fragments are abundant; features on the mesa top include large stone outcrops used as grinding areas. Ceramic assemblages include transitional Glazes E to F (Kotyiti glazes), suggesting an occupation extending from about 1525–1600 to 1700. A. H. Warren's notes on earlier collections document temper materials that include Zia basalt with rhyolite tuff, and black vitrophyre and rhyolite tuff indicating local manufacture of pottery as well as trade with peoples from Zia, traditional trading partners and allies of Jemez peoples. More thorough analysis of temper materials in glaze and other historic Pueblo pottery types is needed to clarify the issue of source materials.

BOLETSAKWA (LA 136, FS 2), LITTLE BOLETSAKWA (LA 135, FS 199), "PLACE OF THE SEASHELL"

Previous Research, Boletsakwa (LA 136). Harrington 1916; Bloom, 1922; Stallings 1933, 1937; Reiter 1938; Smiley 1951, Museum of New Mexico 1968;[11] Elliot 1982.[12]

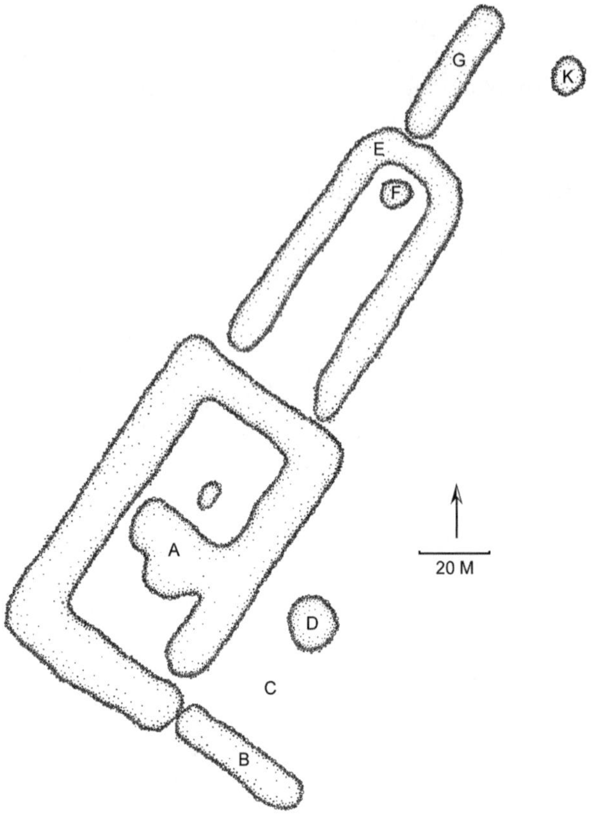

FIGURE 21. Boletsakwa.

Previous Research, Little Boletsakwa (LA 135). Housley 1974.[13]

Site Location, Boletsakwa (LA 136). Boletsakwa is located at the summit of a narrow projection, or *potrero*, just west of a small canyon at an elevation of 7,240 feet (see figure 21). Immediately to the north lies a small creek, and just west of the mesa is a larger riparian stream.

Site Location, Little Boletsakwa (LA 135). Little Boletsakwa (LA 135) is located two hundred meters to the north of the main village (see figure 21). The mesa narrows gradually to a forty-meter-wide section north of the main village. From here there is a well-defined saddle in the mesa. The sharp change in elevation separates the main pueblo from the smaller, fortified habitation area known as "Little Boletsakwa." LA 135 is situated on a slightly elevated, triangular portion of the mesa and is separated from LA 136 by a low (1- to 1.5-meter) wall that extends the length of the southern approach to the community.

The approach to the canyon and the site would have been difficult prior to the construction of Forest Service roads, and the site is largely obscured by the large stands of Ponderosa pines in the area. The mesa top is covered with cholla, and potentially cultivatable areas are available on the summit as well as in the narrow valleys that surround both creeks to the east and west of the habitation area.

Site Description LA 136, Boletsakwa. Boletsakwa is a very large rectangular pueblo situated in a clearing at the summit of the mesa. The site consists of three distinct roomblocks that stretch roughly north to south in a linear arrangement following the shape of the mesa. The largest (roomblock A) is rectangular in shape and contains two central plazas divided by a single linear roomblock measuring 120 × 80 meters. This central habitation area was most likely almost completely enclosed; only a small break in the southeast corner leads to the outside of the roomblock. Based on the size of the rubble mounds, this roomblock may have been two to three rooms deep. In the central bisecting roomblock, the size of the mound (roughly 20 × 40 meters) is indicative of either a series of larger rooms or a three- to four-deep row of individual rooms. The northern, southern, central, and eastern portions of the site have mounded rubble from 1 to 1.5 meters high, suggesting a multiple-storied structure.

Immediately to the west of this main roomblock is a single detached row of rooms measuring forty to fifty meters in length (roomblock B). North of this structure is a circular depression, roughly fifteen meters in diameter (depression C), and to the north of this is a similarly sized kiva (D). The depression lacks any stone masonry and was most likely a reservoir of some kind, although some (Elliot 1982) have identified it as a kiva. The kiva appears to be a semisubterranean feature. Approximately fifteen to seventeen meters in diameter, the structure is circular in shape and has walls exposed by illegal excavations. Both features are located in a plaza area that extends northward along the outer walls of the main roomblock.

North of the main habitation area lies a detached smaller, U-shaped structure (E). The height of the mounds (one meter) and the width of the roomblocks (eight to ten meters) suggests a single-storied roomblock with a single row of rooms. Points of entry are clearly present at the southern, open portion of the U-shaped roomblock, but the northern area is completely enclosed. This single structure surrounds the third plaza, which measures approximately 40 × 80 meters. In the northern section of the plaza, sheltered by the enclosed end of the roomblock, is a small kiva (F). This smaller mounded depression is ten to twelve meters in diameter.

At the northern end of roomblock E is a single row of rooms, approximately 10 × 40 meters in length and similar in mound height to roomblock E. This may have been a single-storied structure but is partially obscured by brush. At the northeast and northwest rims of the mesa lies an extensive row of low defensive walls, constructed out of shaped tuff blocks. These walls are highly eroded and appear to have been no more than one meter in height. There also appear to have been a series of small structures (possibly rooms), facing the northern cliff where the natural saddle separates Boletsakwa from Little Boletsakwa. Clearly, both habitation areas were located with isolation and defense in mind. The building materials and masonry style for the entire site, as revealed by exposed and partially excavated walls are shaped tuff with mortar. Average stone size is roughly 0.75 × 0.50 meter.

Site Description, Little Boletsakwa (LA 135). Little Boletsakwa consists of a large mounded area of rubble. To the north are some exposed areas of bedrock used as a grinding area. The site has been extensively potted, and piles of stones are scattered throughout the small, flat area where the village once stood. The entire site stretches approximately ninety meters from north to south and is about thirty-five to forty meters wide. While there are wall alignments visible in the areas where pot hunting has occurred, there is no apparent pattern to the site. Still, building materials are consistent with other occupation areas at LA 136. The ceramics at the site as well as the grinding stone feature at the northern prominence suggest that the two sites were occupied at the same time. There is a large wall, roughly 1 × 1 meter and 40 meters in length that separate the two occupation areas. If the site were reoccupied at a later date and LA 135 is the older of the two sites, the construction of the defensive wall makes little sense.

Room Number Estimates, Boletsakwa (LA 136). Boletsakwa was a significantly large community in the Jemez Mountain region. While the total number of stories for each roomblock is impossible to know without evidence revealed from excavations (i.e., the placement of hearths, wall thicknesses, the placement of doorways, vents, and viga holes), Boletsakwa is thought to have held 660 to 700 rooms.

Room Number Estimates, Little Boletsakwa (LA 135). While the room unit arrangement is difficult to discern without excavation, there appears to have been a fairly dense habitation at the site. Based on the amount of rubble and the size of the mounds at the center of the site, Little Boletsakwa (LA 135) may have held three hundred to four hundred rooms.

Artifact Descriptions, Boletsakwa (LA 136). The artifact density for Boletsakwa is very high. Basaltic hammerstones, basaltic grinding stones, chalcedony flakes, obsidian flakes, and cores are scattered throughout the pueblo. A midden area extends around the eastern, downslope-facing portion of the mesa. Ceramic densities are similarly high. Santa Fe Black-on-White are noted in an early report, although none were observed at a later visit. Jemez Black-on-White, Tewa Polychromes, and plain and smudged utility wares are present. Glazewares D, E, and F are present as well, suggesting a fairly continuous occupation of the site, from the late 1490s until the late 1690s.

Dendrochronology. Extensive dendrochronological dates exist for Boletsakwa, although the specific provenience of these materials is not known. Cutting dates collected from the site indicate that construction possibly began as early as 1656 ($N=1$) or 1663 ($N=1$) but that the majority of materials collected were cut in the years immediately following the Revolt: 1680 ($N=3$), 1681 ($N=2$), 1682 ($N=1$), and 1683 ($N=5$) (Elliott 1982: 87). Obviously, some of the earlier material could have been salvaged from earlier construction projects, or it may have been collected from fallen timbers. Interesting as well are the later cutting dates. Boletsakwa was a large pueblo (660+ rooms at the main village and 350+ at Little Boletsakwa), and construction seems to have proceeded through the years immediately following the Revolt, at least until 1683. Observations of ceramics at Little Boletsakwa (mostly Jemez Black-on-White) suggest that this component of the site was occupied at an earlier date. Without the specific provenience of the dendrochronological dates, it is impossible to know whether sections of the two sites were constructed or occupied at the same time.

Artifacts and Ceramic Descriptions, Little Boletsakwa. The presence of this broad spectrum of utility wares, hammerstones, flakes, and decorated pottery indicates that Little Boletsakwa was definitely a substantial habitation area. Ceramics from this site consist of mainly Jemez Black-on-White and corrugated, smudged, and plain utility wares, although unidentified glazeware fragments are present in the eroded portions of the site. Hammerstones, obsidian and chalcedony flakes, as well as a variety of grinding stones are present at the site, although considerable erosion has occurred on the three sloping areas of the mesa.

Occupation Period LA 136, Boletsakwa. While the dendrochronological dates are fairly straightforward in indicating a Revolt period construction phase (1156–1683), the ceramics collected suggest a much longer occupation sequence for the site, possible from the 1490s to the 1690s. Stallings's work does not specify where the materials were collected from. It now appears that roomblocks A to F date specifically to the Revolt period. Because of the close proximity of the two villages, the two habitation areas can and have been identified as a single site.

Occupation Period LA 135, Little Boletsakwa. On the basis of ceramic information, a clear occupation period for LA 135 is difficult to establish. Certainly, the Jemez Black-on-White sequence includes the more clearly defined occupation sequence of LA 136. However, the construction of a defensive fortification at the southern end of the site, at the crest of the saddle and between the two habitation areas as well as the inaccessible position of the habitation area, suggests that Little Boletsakwa was built with defense and isolation in mind. Still, it seems likely that Little Boletsakwa was the result of an earlier occupation (ca.) 1350–1625.

SESHUKWA OR SAN JUAN MESA RUINS, "EAGLE DWELLING PLACE" (LA 303) (FS 12)

Previous Research. Bloom 1922; Reiter 1938; Smiley, Stubbs, and Bannister 1953; Elliot 1982; Raish 1990.

Location. Seshukwa is located at 7,900 feet on the eastern edge of a narrow projection of a large mesa in the Jemez region. The site is situated in a broad clearing of Ponderosa pine association forest. The approach to the village is treacherous; the steeply inclined cliff faces drop nearly 150 meters to the river valley below.

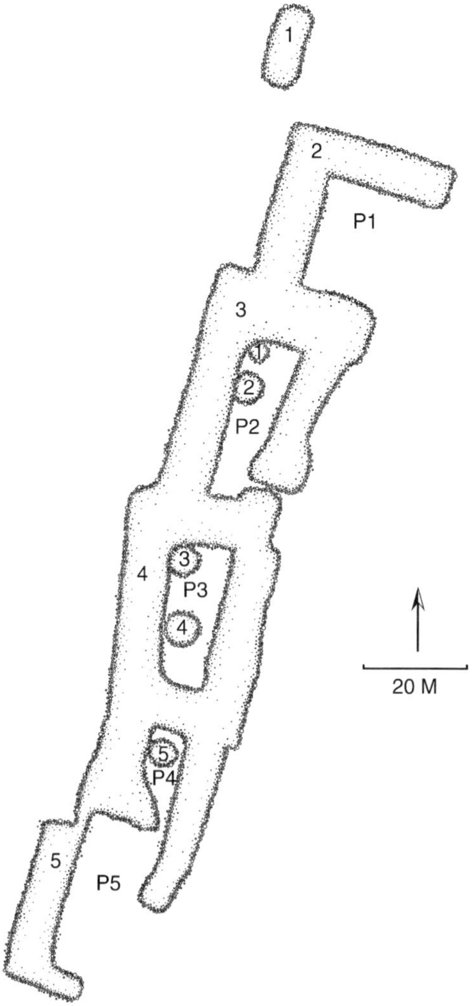

FIGURE 22. Seshukwa.

Site Description. Seshukwa is a large site, containing as many as one thousand rooms (see figure 22). Five contiguous roomblocks, five plaza areas, and six kivas are present; an isolated kiva lies 100 meters south of the main village. Seshukwa is constructed of large shaped tuff masonry blocks. The village stretches 210 meters north to south and covers 40 to 50 meters east to west along the mesa edge. Fifty-seven rooms are clearly visible on the surface; other visible wall alignments reveal roomblocks four to eight rooms deep. The site has suffered from extensive pothunting damage. Still, the standing walls still visible throughout the site (some as high as four meters) suggest that the village was a multiple storied structure. Several small single- and double-room structures, often referred to as field houses, are present throughout the mesa top. These have not yet been systematically mapped.

Beginning at the extreme north is roomblock 1. This compact structure measures 15 × 8 meters and is slightly offset from the main village. Ten meters southeast lies roomblock 2. Measuring 35 × 10 meters, this feature has several exposed, roughly square rooms measuring 3 × 3.5 meters each. The southern set of rooms faces plaza 1. Plaza 1 is 30 × 30 meters and opens to the eastern cliff face. Just south is the main section of the pueblo, a continuous series of parallel structures enclosing three sheltered plaza areas and one (plaza 5) that opens to the east.

Roomblock 3 has a northern section that faces plaza 1 and continues south for 60 meters. This rectangular segment is fairly broad (35 × 15 meters) and suggests a multiple-storied construction. Located in the center of the roomblock is plaza 2. This completely sheltered plaza encircles two small kivas. Like many of the Jemez sites, these similarly sized (six to seven meters in diameter) kivas are set in the northwest corner of the plaza. In the southwest corner is a possible entrance. Connected to the southwest section of roomblock 3 is roomblock 4. Longer (seventy meters) and slightly wider (thirty meters) than roomblock 3, roomblock 4 is a rectangular structure containing two small enclosed plaza areas. Plaza 3 frames two kivas (K3–K4) situated in the northwest corner of the feature. Kiva 4 is slightly larger (ten meters in diameter) and occupies a central position in the plaza. South of plaza 3 is a small projection of rooms separating plaza 3 from plaza 4. Plaza 4 is considerably smaller (20 × 10 meters) and is completely enclosed by the a multiple-storied section to the west and an apparent single-storied section of roomblock 4 to the east. Southeast of roomblock 4 is roomblock 5. Roomblock 5 appears to have been a discontinuous section of the village. Measuring 30 × 15 meters, roomblock 5 contains a series of small (3 × 3-meter) rooms surrounding a much larger square-shaped structure, kiva 5. The entire roomblock surrounds plaza 5, which opens to the southeast.

Despite the high elevation of the site, Seshukwa appears to have been the site of extensive agricultural production. The immediate mesa top area contains several potentially cultivatable field sites, both north and south of the main village. Below the pueblo, on lower sections of the mesa, are several other areas that may have served as field sites.

Intermittent streams crosscut the mesa, and streams, fed by runoff, flow through the valleys to the east and west of the site.

Room Number Estimates. The amount of mounded rubble and the height of standing walls suggests that Seshukwa was a multiple-storied structure. Rough estimates of individual sections are roomblock 1, 30 rooms; roomblock 2, 50 to 75 rooms; roomblock 3, 250 to 350 rooms; roomblock 4, 400 to 500 rooms; roomblock 5, 50 to 75 rooms. This yields a total estimate of 780 to 1,030 rooms.

Artifacts and Ceramic Descriptions. Artifact density at the site is extremely dense. Ceramics present include Jemez Black-on-White, Glazewares D to F, and corrugated and smudged utility wares. Lithics present include vesicular basalt and sandstone manos and metates, obsidian, and chalcedony flakes and cores.

KWASTIUKWA, "GIANT FOOTPRINT RUIN" (LA 482, FS 11)

Previous Research. Bandelier 1892: 205–206; Reiter 1938: 20; Elliot 1982: 23.[14]

Location. The site is located at 7,600 feet on a narrow projection of a large mesa in the Jemez region (see figure 23). A smaller site (FS 7, LA 483) is located two miles to the north on the same mesa. Ceramics from this smaller (three hundred rooms) site include Jemez Black-on-White, Glazes C, D, and E. Pecos Glaze E Polychromes have a range of 1515 to 1700 AD and indicate parallel or contemporary occupations of these two sites, perhaps into the Revolt periods.

Site Description. This poorly known site may be one of the largest yet recorded in the Jemez region. Building materials are varied and include shaped and unshaped volcanic tuff as well as adobe. The site has a fairly complicated architectural pattern and consists of seven large roomblocks (labeled A–F in figure 23) arranged around seven plaza areas (P1–P7). Five kivas are present within the immediate vicinity of the pueblo, one of which is of sufficient size to be considered a great kiva. There is a smaller kiva located approximately one hundred meters east of the main site as well. There is also a large reservoir in the extreme northwest portion of the site.

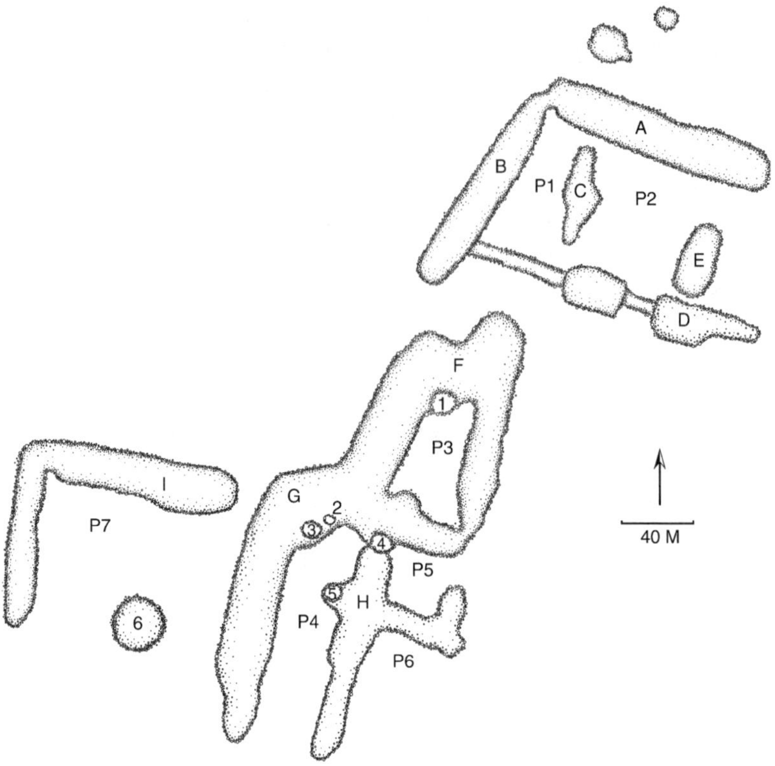

FIGURE 23. Kwastiukwa.

The village stands in the midst of a secondary Ponderosa pine forest although the main habitation area has remained free of large vegetation. There are large areas of cultivatable lands and fertile soils in the immediate vicinity of the village. Semipermanent streams flow to the east and the west of the main habitation areas and may have allowed limited water diversion irrigation. There are also permanent rivers and streams in the river valleys to the east, west, and south of this community.

The roomblocks can be separated into three discontinuous units. The northernmost set of roomblocks (A, B, and D) form a single, open-ended rectangle. Individual roomblocks are oriented roughly north-south and east-west. Of particular interest is roomblock A, a section of 100 to 120 ground-level rooms known as "the barracks." Roughly twenty to thirty of these rooms have been disturbed by previous excavations, and several of the rooms have exposed, datable roof beams. It has been assumed by some researchers (Elliot 1982: 22) that this portion of the site was temporarily reoccupied during the Revolt period, although the exposed rooms may have been recently excavated by pot hunters. These exposed rooms, measuring approximately 3 × 8 meters and standing in rows four to five deep, give an indication of a general size and orientation of individual rooms. These dimensions, coupled with the height of the rubble (one to three meters), suggest that the majority of the roomblocks supported at least two and in some cases as many as three to four levels.[15]

To the north of roomblock A are two smaller clusters of twenty to twenty-five ground floor–level rooms. Immediately to the west of these structures is a substantial (25 × 25-meter) reservoir. A low embankment, open to the east, has been constructed around the perimeter.

Roomblock B is a section of 100 to 125 ground-level rooms and is joined at its northeast corner to roomblock A and at the southeast corner to roomblock D. Roomblock D is less substantial then roomblocks A and B, but large mounds at the eastern end of the building are considerably higher and suggest a multiple-storied section of rooms. All three of the main roomblocks form a large central plaza measuring 120 × 100 meters. Roomblocks C and D each contain twenty to forty ground-level rooms, and split the plaza into two distinct areas (labeled P1 and P2).

The second, discrete habitation area (roomblocks F, G, and H) lies twenty meters to the south of roomblock D. Roomblock F is rectangular in shape and is oriented along a similar north-south grid, yet it appears to be a substantially more compact structure than the other habitation areas. Measuring roughly 70 × 90 meters, room depth from the outer walls to the small plaza (P3) most likely range from four to seven rooms. In the northwest corner of plaza 3 is a circular kiva, roughly ten meters in diameter. The eastern section of roomblocks contains the partially excavated remains of fifteen to twenty rooms.[16] The western and southern roomblocks are heavily disturbed by illegal pot hunting but are of substantial size and depth. Based on the height of the rubble mounds (one to three or more meters) and the size of the exposed rooms, the total number of ground floor–level rooms is estimated at 400 to 425.

The southwestern corner of roomblock F is joined at a diagonal angle with a similarly sized structure, labeled roomblock G. Roomblock G has a similar north-south orientation and continues west for twenty meters in a roughly a diagonal angle and continues south for approximately one hundred meters. With roomblock H, located forty meters to the east, this structure helps frame the fourth plaza area (P4). Plaza 4 is long and rectangular, is open only at the southern end, and measures 20 × 80 meters. At the extreme northeast of plaza 4 (in an arrangement similar to plaza 3) is kiva 2, a structure of similar size and shape (ten meters in diameter) to kiva 1. Roomblock G is heavily disturbed by illegal excavations as well as a large bulldozer cut in the southern portion of the structure. Ground floor room estimates range from 200 to 225. This roomblock appears to have supported multiple stories.

Plaza 4 is enclosed by the fifth major structure, roomblock H. Roomblock H is a T-shaped structure, containing approximately seventy-five to eighty ground-level rooms. Room depths, from plaza to plaza, are estimated at six to eight rooms at the central axis and taper to a depth of two to three rooms at the southern and northern extremes. The central segment may have supported multiple stories. The two smaller northern projections frame the fifth plaza area (P5). The northern portion of roomblock H is possibly joined to the southern portion of roomblock F. In a pattern similar to plazas 3 and 4, plaza 5 (measuring 20 × 20 meters) contains a small kiva in the extreme northeast corner. The eastern extension of roomblock H frames a larger plaza (P6). This area is open to the southeast and measures 40 × 30 meters.

The final discrete habitation are is roomblock I. This L-shaped structure consists of two sections; the northern segment runs east-west for seventy to eighty meters, and the second segment runs north-south for sixty meters. These two segments frame the seventh plaza (P7) and encloses a centrally located great kiva. This roomblock is significantly more narrow than roomblock F and may have supported multiple stories as well. The great kiva is mounded on all sides and measures fifteen to twenty meters in diameter. It is not clear whether this structure supported a roof.

The arrangement of all three habitation areas does follow a pattern; roomblocks A, B, F, G, and I segregate the plazas from the eastern approach and may have been constructed with defense in mind. The site is also surrounded by several small one- to two-room sites. While a great deal of research has been devoted to the function of isolated single- and double-room field houses, little is known about the function of similar structures located in close proximity to large Revolt and contact period sites. Damage to the site from earlier excavations is extensive. Pot hunting has likewise contributed to the deterioration of the site. A Forest Service logging road has been cut directly through the center of the village, and a fairly recent bulldozer cut has been made through the center of the southern portion of roomblock F.

Room Number Estimates. Earlier studies of Kwastiukwa suggested that the site was one of the larger habitations in the Jemez Region. Elliot estimated the total number of rooms at 1,250 (1982: 22). More recent studies of the site, initiated by Whatley (discussed earlier), have determined that Kwastiukwa was a much more substantial, multiple-storied settlement containing as many as 2,500 to 3,000 rooms. The estimates of individual roomblocks A to D contain between 430 and 450 ground-level rooms. Roomblocks F, G, and H may contain as many as 400 to 450 additional ground-level rooms, and roomblock I may have contained as many as 150 ground-level rooms. This total of about 1,000 ground-level rooms, coupled with evidence supporting multiple stories in most areas of the site, yields an estimated total of 1,700 to 2,300 rooms. This would qualify Kwastiukwa one of the largest historic Puebloan sites on record. Surprisingly, no mention is made of a site of this size (located in the Jemez province) in Spanish historical documents.

Artifacts and Ceramic Descriptions. Artifact density is extremely high for the site, although a systematic (or random) sampling of individual roomblocks has not yet been conducted. This raises questions about the various occupation periods of the site; without an assessment of construction sequences, room totals for any specific occupation period can only be estimated. Jemez Black-on-White, Glazes D, E, and F, are found throughout the site. Plain and corrugated utility wares are also found in abundance. Lithics are found in high densities and include chert, chalcedony, and obsidian flakes and cores. Basaltic manos and metate fragments are also scattered throughout the site. Several exposed roof beams are present throughout the site, but no known samples have been collected. There has also been no systematic survey of the mesa.

AMOXIUMQUA (LA 481, FS 530)[17]

Previous Research[18]. Bandelier 1890: 126, 1892: 205–206, 1975: 225, 213; Holmes 1905: 206–208; Hewett 1906: 48;[19] Chapman 1911; Harrington 1916: 395; Hodge 1905: 242; Stallings 1937: 5; Reiter 1938; Smiley 1951; Smiley et al. 1953.

Location. Located at the summit of a large mesa in the Jemez region (7,800 feet), the community is perched on an elevated section of the mesa.

Site Description. Amoxiumqua is a large, multistoried masonry structure covering approximately twenty to twenty-five acres (see figure 24). According to Hewett's measurements (1906: 49), the entire arrangement of visible features measures 350 yards (northeast to southwest), reaching a width of 200 yards in some places. The site is located approximately four hundred meters from the western rim of San Diego Canyon in a

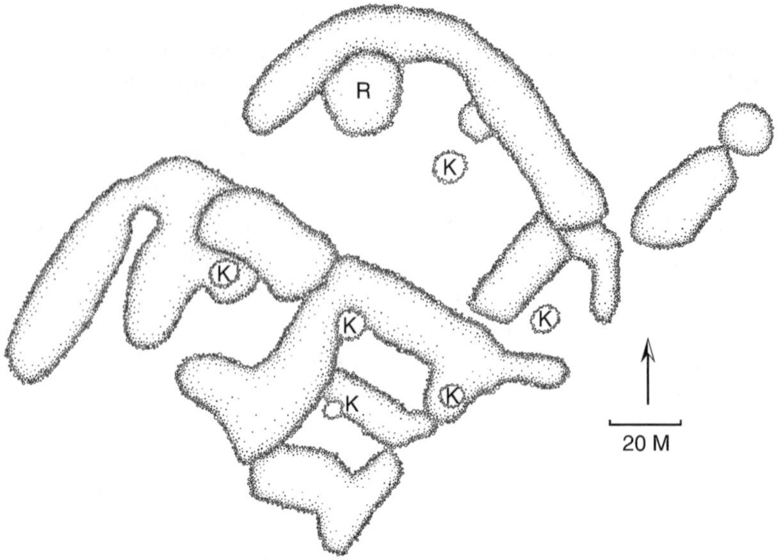

FIGURE 24. Amoxiumqua.

clearing of Ponderosa pine forest. Jemez Creek in the valley below is barely visible from the outer walls of the village, but the community is difficult, if not impossible, to see from the valley floor.

One of the larger villages in the region, the layout of the site suggests a rambling arrangement of roughly rectilinear roomblocks arranged around four large plazas. Two of the plazas are at significantly higher elevations than the others. A total of nine circular kivas are present, most of which measure approximately twenty-five feet in diameter. A single larger kiva measures thirty-two feet in diameter. Room sizes are generally larger on average than those at other contemporaneous Jemez sites (no quantitative data were collected in the excavation notes), and a large, semicircular earthen dam is located to the north of the main settlement. In 1905, Holmes (207) estimated its diameter at 150 × 6 feet deep, although considerable erosion and disturbance have occurred since then. Potentially cultivatable lands are present on the top of the mesa as well as in an area located to the east of the main settlement.

The construction materials are also quite different than Unshagi or Nanishagi. Masonry materials are well-shaped, regularly sized tuff, breccia, and sandstone; and both adobe and masonry walls are present at the site. Wall thicknesses are relatively greater than those found at other sites, and the overall arrangement of the individual walls (many of which are at right angles) suggests a more carefully planned, less harried construction sequence. The walls were apparently replastered at least three times as well. Lacking from any of the notes are a possible construction sequence; wall joins were apparently not carefully noted, despite the many seasons of excavation. The main roomblocks each consisted of three to four stories. At the time of Hewett's publication (1906), walls were still completely visible, and roof beams and other organic materials were also present. A number of small one- to four-room structures are scattered in the immediate vicinity of the site. Elliot suggested that these structures are field houses, although the possibility exists that these may have housed people who could not be accommodated at the main settlement. Rock art located near the cliff (FS 577) suggests some kind of trail marker or may represent a shrine.

Room Number Estimates. 1,250 to 1,500 rooms.

Artifacts and Ceramic Descriptions. Jemez Black-on-White, numerous uncategorized plain and corrugated utility wares, as well as Glazes A to F are present at the site. Lithic fragments of obsidian, chalcedony, and red sandstone and rhyolitic basalt groundstone, mano, and metate fragments are present as well. Unusual artifacts noted in the excavation notes include an object referred to as a "flesher" shaped from the distal end of an adult femur. Much of the excavation centered on the collection of burials and burial materials. In most cases, rooms were cleared to reveal burials (130 are mentioned) and burial goods. "Venetian" beads found with two juveniles support a historic association of the site. Tree ring materials collected by Stallings (1937: 5) reveal a single cutting date of 1502, although the provenience of the wood is not clear. These materials, as well as the presence of Glaze F (1650–1700) sherds, indicate that the site was occupied from 1502 to at least the time of the Revolt. Bandelier (1890: 204–205) argues that the site was abandoned prior to 1680 but provides little supporting evidence.

TOVAKWA (LA 483) (FS 576)[20]

Previous Research. Harrington 1916; Reiter 1938, Elliot 1982; Ramenofsky, Penman, and Steffen 1996.

Location. The site is situated at 7,900 feet, near the face of steep cliff on the eastern section of a large mesa. To the west is a steep canyon with a permanent stream fed by a series of small springs. The village stands in the midst of a Ponderosa pine forest. Vegetation throughout the site is dense. The mesa top has several open, flat areas with sufficient soil and runoff to support dry farming. The canyon lands at the base of the mesa have several areas that would have been suitable for floodplain or irrigation farming.

Site Description. This site is one of the largest in the Jemez region and contains as many as 1,800 to 2,000 rooms (see figure 25). The entire area of the village has been estimated at twenty-five acres. Extending nearly 400 meters north to south and 350 meters east to west, the sprawling complex of roomblocks curves along the steep, eastern cliff above a narrow river canyon 800 feet below. The main portion of the village was built on a flattened area of the mesa, although the difference in elevation between the northwest and the southwest areas of the site is nearly twenty meters. The site is framed by two arroyos, one to the southwest and the other on the southeast border of the site. Construction materials are shaped and unshaped volcanic tuff.

The main village is roughly rectangular in shape and consists of a series of five discontinuous roomblocks, twelve small, round (ten meters in diameter) kivas (four of which are freestanding), one great kiva, and six plaza areas. The northernmost roomblock, often referred to as the "barracks," is a small rectangular section of twelve ground floor–level rooms located approximately one hundred meters north of the main village. Wood fragments (possibly the remnants of roof beams) and collapsed walls are still visible. The structure measures thirty by ten meters. To the west of the barracks is a Forest Service logging road that leads from the northern summit of the mesa to the village. Running directly through the center of the site, the road may have been built on the remnants of a trail connecting the barracks to the main group.

Roomblock 1 is an L-shaped feature that faces the steep incline of the eastern cliff. Roomblock 1 has two segments. The main structure measures one hundred meters east to west and stands twenty to twenty-five meters wide. The second, southern projection of rooms measures 40 × 15 meters. Portions of roomblock 1 may have been two stories high. Although the wall and room alignments are impossible to determine without excavation, room number estimates range from 150 to 200. Between the southern side of the roomblock and the cliff is a small, freestanding kiva measuring ten meters in diameter.

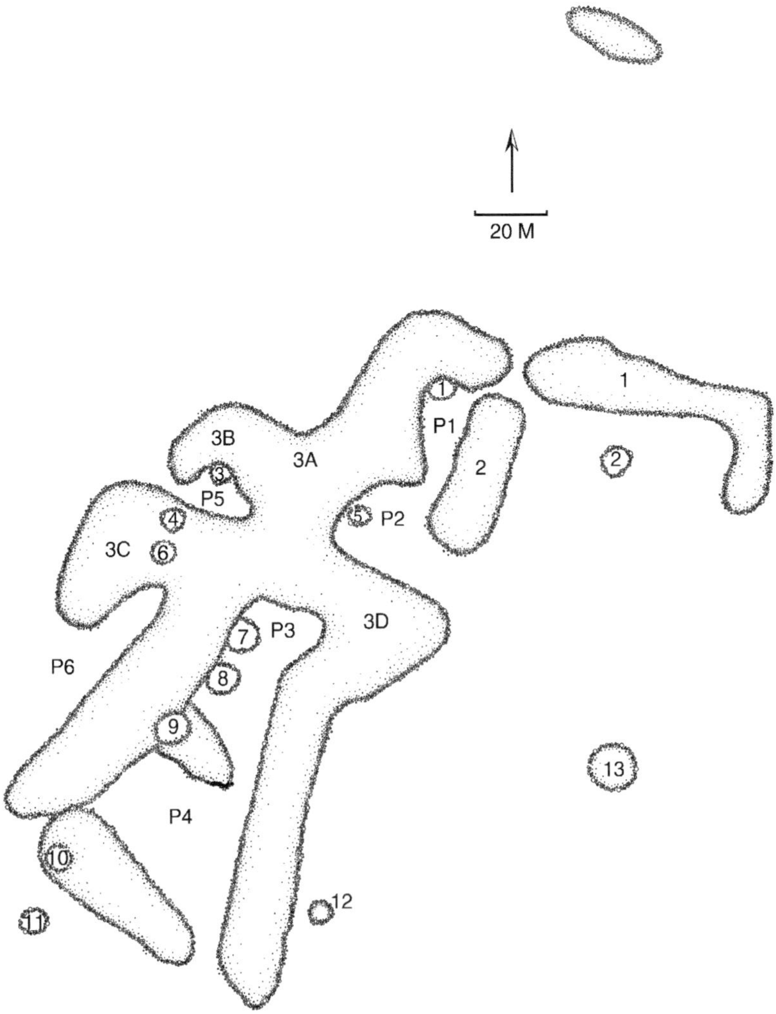

FIGURE 25. Tovakwa.

Immediately to the west is the northern section of roomblock 2. This rectangular group of rooms runs north to south and measures 70 × 20 meters. To the west of the structure are two small isolated plaza areas, bisected by a projection of rooms from the center of roomblock 2. The structure was probably multistoried and may have contained 150 to 200 rooms.

Plaza 1 measures approximately 20 × 40 meters and contains a small kiva (kiva 1) built into the eastern face of roomblock 3. Directly to the southwest is the slightly larger (35 × 30-meter) plaza 2, and in the western corner is a freestanding kiva (kiva 5). Both plazas are enclosed by the central east-west section of roomblock 3.

Roomblock 3 is a complex structure and can be subdivided into four continuous sections (labeled 3A–3D). Section 3A is a continuous, fairly wide (20- to 30-meter) multistoried structure stretching 230 meters northeast to southwest. Moving from north to south, section 3B is an L-shaped section measuring 20 × 30 meters. The sections enclose a small (20 × 20-meter) plaza (P5). Built into section 3B is a small kiva (K3), similarly located in the northwest area of the plaza.

Section 3C similarly extends to the west in an L shape. Measuring 50 × 40 meters, section 3C contains two small kivas (K4 and K5). Plaza 6 opens to the southwest, and one of the kivas (K5) appears to have been situated in the northwest corner of plaza 6.[21] Directly opposite section 3C is section 3D. Extending southeast for 50 meters, this long, L-shaped section runs along the eastern cliff face for 140 meters and has a depth of 20 to 30 meters.

The open area between sections 3D and 3A can be divided into two separate plaza areas (P3 and P4). Two kivas (K7 and K8) are situated in the northwest area of plaza 3. Kiva 7 is built into the eastern face of section 3A. Kiva 8 is a freestanding structure of similar size, five meters to the south. Separating the two plazas is a short section of rooms attached to section 3A. Located close to the northwest section of plaza 4 is a small kiva (K9), possibly built into the roomblock. Plaza 4 is the largest at the site (50 × 70 meters).

To the south lies roomblock 4, a separate structure measuring 70 × 20 meters. Two small kivas are present in this southern section of the village. The first (K10) is set into the roomblock, and the second is set ten meters to the southwest. A seventh plaza may have existed at the southern extreme of the village. If so, these two kivas would have stood in the same northwest area of the plaza. Perhaps the most unusual feature at the site is a large great kiva (fifteen to twenty meters in diameter) perched on an outcropping of volcanic tuff thirty meters down the eastern face of the mesa.

Clearly, Tovakwa was built to minimize access to the central portions of the village. Based on the elevation of the site (7,800 feet), coupled with the location of the roomblocks along the eastern edge of the mesa, defense was probably an important consideration in the construction of this community. The only clearly discernible entrance to the main village is in the southwest corner of plaza 4. Other possible entry points are in the western area of plaza 5 and in the gap created by the Forest Service road between the northern section of roomblock 3 and the western section of roomblock 1.

Room Number Estimates. Elliot classified Tovakwa as the largest site in the Jemez area, estimating a total of 1850 rooms for this community (1982: 56). Individual roomblocks are obscured by thick vegetation, and only rough estimates can be made for each of the roomblock sections. Based on the dimensions of each section and the height of rubble, the following estimates seem probable: roomblock 1, 200 rooms; roomblock 2, 150 to 200 rooms; roomblock 3A, 600 rooms; roomblock 3B, 50 to 75 rooms; roomblock 3C, 75 to 100 rooms, roomblock 3D, 200 to 250 rooms; roomblock 4, 100 to 150 rooms—for a total of 1,425 to 1,575 rooms.

Artifacts and Ceramic Descriptions. Artifact density at the site is extremely high. Jemez Black-on-White; Glazewares D, E, and F; and plain and smudged utility wares are present throughout the site. Vesicular basalt and red sandstone manos and metates were utilized as well. Chert, chalcedony, and obsidian cores and flakes are likewise present.

TAMAYA (LA 2049), "CANJILLON PUEBLO" D2 "RUIN"

Previous Research. Bandelier 1892: 193–196;[22] Mera 1940a: 26; T. Williamson Museum of New Mexico site card 1937; Warren 1970.

Location. Tamaya is located at approximately 5,830 feet near the summit of a large mesa in the Jemez region. The village is approximately ten miles north of the San Felipe and Santa Ana pueblos. Perched on the southern edge of a peninsular projection, ninety meters above the canyon floor (see figure 26), Tamaya is situated near the confluence of the Rio Grande and Jemez rivers. The site is accessible from the north only and is separated from the rest of the mesa by a lower, broken cliff face twenty meters in height. Partially concealed basaltic stone fortifications are scattered along the trail to the mesa, suggesting that defense was an important consideration in the construction and location of the site. The lack of appropriate areas for cultivation and water sources at the summit supports this interpretation.

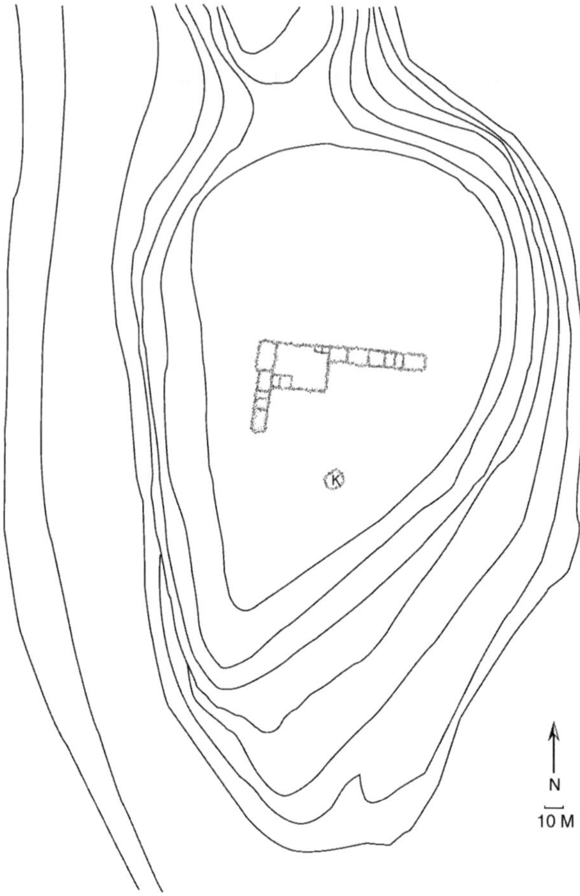

FIGURE 26. Tamaya (Cerro Colorado).

Site Description. Tamaya consists of an L- or V-shaped room block arrangement, located one hundred meters from the southern edge of the peninsular, vesicular basalt formation, with a plaza in the interior of the angle. The larger, northern room block has an east-west alignment and measures forty-five to fifty meters end to end and contains approximately twelve to twenty-four variably sized rooms. Wall heights are approximately 3 to 4 feet in places, but rubble heights of 1 to 1.5 meters most likely suggest a single-storied structure; wall thickness is between 25 and 50 centimeters. Building materials include shaped stone blocks measuring roughly 20 × 50 cm. Room sizes are variable (between 2 × 3 and 7 × 5 meters), with four smaller (2 × 2-meter) rooms located in the interior of a larger, 10 × 12-meter wall alignment (at the center of the angle). The purpose of this unusual feature is unknown but could represent a partially constructed roomblock abandoned prior to completion. There is also evidence that the site was burned, possibly by Spanish conquistadors, following the rebellion of 1680.

The southern, north-south-oriented room block is smaller, measuring approximately 20 to 25 meters and includes four, more regularly shaped (7 × 4-meter) rooms. To the south of the angle is a shallow depression, most likely a kiva, carved out of the basaltic bedrock, approximately five meters in diameter. Sherds collected from the immediate area include those of an earlier style (Kana'as Banded and Lino Gray, characteristic of a P1 occupation), but mounded materials from the shallow excavation are still present near the depression, indicating a more recent and hasty Revolt period construction.

Room Number Estimates. Twenty-five to fifty rooms of irregular size and shape.

Artifacts and Ceramic Descriptions. Mera's (1940a: 26) description of the site is fairly sparse, but he suggests that "this site appears to have been inhabited solely during the production of (Glaze) Groups 'E' and F.'" Mera's sample collection contains few diagnostic rims, and the majority of the collection contains undecorated sherds. Samples collected from the site by in 1937 by T. Williamson contain Lino Gray Plain and Kana'as Banded sherds, suggesting either an earlier Pueblo I (700–1100 AD) or developmental Pueblo (600–1200 AD) occupation. Alternatively, the kiva may be associated with a site located at the southern end of the mesa. Type collections on file at the Museum of New Mexico Lab of Anthropology contain Whitemound and Red Mesa Redwares, Trenaquel Glaze Polychromes, Kapo Black, and San Ildefonso Red polished wares as well as a smaller number of Tewa Polychrome fragments. Each is roughly indicative of a mid– to late seventeenth–century occupation of the site.

Subsequent reanalysis by A. H. Warren (1970: 3) reveals a more diverse array of ceramic types present at the site: "plain undecorated sherds predominate; forms seem to be bowls and ollas. . . . Utility wares are of the Santa Cruz type with polished interiors and flaring rims." Many seem to be slipped on the inside, as do the Cochiti utility wares of the late seventeenth century. More relevant to the discussion of the rejection of Spanish material culture and ideology is the apparent absence of any type of Spanish pottery (characterized by the "soup plate" forms) at the site. In both plain wares and utility wares, temper materials are noted as "fine grained light to medium gray basalt, most probably of local origin."[23] Warren suggests that based on stylistic similarities and temper materials, the pottery at the site was most likely produced at San Felipe. According to Warren, temper in the glazewares is similar in appearance to the tempers used in the Cochiti area.

Lithic materials are present at the site but in lesser quantities. Chipped chalcedony fragments, flakes, cores, and other debitage are noted in site reports; "Glassy basalt," chert and obsidian fragments, shaped vesicular basalt fragments, and numerous hammerstones are also present and suggest the continued use of Indigenous technologies at the site.

Conclusions

Several conclusions can be drawn from the Jemez region settlements. Based on the centrality of the Jemez region in Puebloan resistance efforts, it appears that several of the ancestral Jemez villages were periodically reoccupied during the several episodes of open rebellion. As mentioned previously, both aggregation and dispersal are evidenced by the construction of large villages. Cerro Colorado, Patokwa, Astialakwa, and Boletsakwa were apparently constructed during Revolt period and housed the majority of the Jemez population. Room estimates from these settlements total 2,250, suggesting that substantial numbers of Jemez people were present at these villages. Ancestral villages such as Seshukwa, Kwastiukwa, Kiatsukwa, Amoxiumqua, and Tovakwa were also periodically inhabited throughout the seventeenth century. Estimates of relative populations at these large villages are impossible to reconstruct without excavations, but it is likely that individual families or clans retired to their ancestral villages during times of social conflict.

Several explanations for these "dispersal movements" are possible. These more remote ancestral villages could have served as staging or planning areas for the Jemez and their allies. Based on the historical documents of the entrada and early colonial periods, it is known that allegiances between the Navajos, the Jemez, and other Pueblos presented serious obstacles to Spanish missionaries and colonists. It is possible that these sites were inhabited by the Navajo allies of the Jemez. Alternately, these settlements could have been inhabited by people who did not support the activities of the pan-Puebloan movement. In either case, a strong argument can be made that the desire to maintain social

TABLE 6.6. Jemez Region Populations Estimates

Date	Expedition	Population Size	Estimated Number of Villages	References
1541	Coronado	n/a	7	Hammond and Rey 1940:259
1581	Chamuscado	n/a	15	Hammond and Rey 1966:107
1583	Espejo	30,000	7	Hammond and Rey 1966:223–224
1598	Onate		11	Hammond and Rey 1953:322
1622	Zarate	6,566	2	Zarate de Salmeron 1966:26
1629	Benavides	3,000	2	Ayer 1916:25
1642	None listed	1,860	1	Scholes 1929:48
1680	Ventancurt	5,000	5	Ventancurt 1698:100
1706	Alvarez	300+	1	Hackett 1923:376
1752	General Census of New Mexico	207	1	Simmons 1979:185

distances or conceal cultural practices from the colonists led directly to their resettlement.

A separate conclusion can be drawn from the Jemez data. Collectively, the large population estimates made by early explorers and colonists are not supported by the archaeological evidence (see table 6.6). The highly inflated figures of the entrada and early colonial period must be considered within their historical contexts. There is little evidence that Coronado, Chamuscado, or Espejo ever visited all (or even any) of the Jemez pueblos.

Espejo's high number is suspect as well. Not only does Espejo not mention having visited each of the Jemez villages, as mentioned in chapter 5, but he asked and was granted permission by the king to make rough estimates of all of the Pueblo populations. These facts are often overlooked by archaeologists and historians who use Spanish population estimates to support massive depopulation of the Pueblo region. In fact, primary sources do not mention epidemic diseases until the mid- to late 1600s, and many of these "famines" were the direct result of Spanish policies. Droughts were frequent occurrences in the Pueblo world. Through several centuries of experimentation, experience, and innovation, the Pueblos had developed survival strategies (and storage facilities) to ensure survival in a marginal environment. Collected as "insurance policies" against droughts, Pueblo surpluses were confiscated through an oppressive tribute system. Adding to the shortages, Pueblo labor was frequently diverted away from the maintenance of their farms to support the economic activities of the colonists. When combined with the droughts of the late seventeenth century, these human-generated famines compromised the health of the Pueblos. Explanations of population crashes that suggest that European diseases alone contributed to the depopulation of the Pueblo region are incomplete.

The estimates themselves are similarly suspect. Selectively drawn from historical sources, estimates of regional population reductions are frequently attributed to diseases by archaeologists. In "mining" these numbers, archaeologists and historians have most likely overstated the impact of disease and underestimated the importance of regional abandonments as a strategy of resistance. The consequences of these two processes are very different. Diseases claim the elderly and young. Elders are the repositories of cultural, religious, and historical memories, and their departure often signals a rupture between past and present historical memories. Abandonments imply the relocation of populations to preserve and maintain social distances between colonists and Pueblos.

Notwithstanding these limitations, the sources of seventeenth-century population figures are questionable for a different set of reasons. Drawn from baptismal and funerary records, these numbers are not census figures in a contemporary sense. Baptismal and funerary records reflect the participation of the Pueblos in Catholic sacraments. Given the tempestuous relationship between the Jemez and their missionaries (five missions were built and burned by the Jemez during the seventeenth century), it is highly unlikely that participation in Catholic sacraments can serve as an accurate proxy for the entire Jemez community. Many communities were split between Christian and non-Christian Pueblos, and it is possible that only Christianized Indians were recorded in these documents.

A separate issue is raised by the relationship between the Jemez and Navajos. These two groups had very fluid social relationships during the colonial period. Because Navajo and Apache sites are more difficult to identify archaeologically, they are most likely underrepresented in archaeological studies. Until we have better information regarding the connections between these two groups and are better able to interpret the numerous, small habitation sites found throughout the Jemez Region, interpretations of large-scale, disease-motivated population crashes should be regarded with caution.

CHAPTER 7

The Archaeological Correlates of Ethnogenesis

Community Building at Old Cochiti

Beginning in the early 1990s, rumblings could be heard within the archaeological community that Native American activists threatened to single-handedly destroy the field of archaeology. Angered over the display of human remains, pot hunting, the desecration of burials, the refusal of museums to grant access to collections by descendant Native communities, and the unwillingness of many archaeologists to consult with Native peoples before conducting research, Native peoples began to lobby Senator Daniel Inouye and President George H. W. Bush for greater control over their ancestral remains. In response, archaeologists quickly mobilized to prevent passage of NAGPRA. For almost a century, archaeologists had enjoyed free access and control over archaeological sites, human remains, artifacts, and, most importantly, the interpretation of those materials.

I was fortunate enough to work with one of a handful of archaeologists who not only saw the legislation coming but also supported the changes it generated—and who, more importantly, was interested in working with contemporary Native peoples. As a young professor at Harvard University, Bob Preucel had seen my application to the graduate program and expressed interest in meeting me and discussing my plans. I had worked in human osteology with physical anthropologist Phillip Walker at the University of California, Santa Barbara. At Harvard, the Archaeology Wing of the Anthropology Department is housed in the Peabody Museum of Archaeology and Ethnology. The department has three wings: Biological Anthropology, Cultural Anthropology, and

Archaeology. If you notice the bird metaphor, you will also notice that a three-winged bird can't possibly fly. "Unusual" and "conservative" is how some people have described the three-winged bird; the Archaeology Wing is no exception. The wing is unique in that unlike many other archaeology departments, Harvard's has never really been identified with any particular theoretical movement (except collecting masses of materials from around the globe). The tectonic shift toward the New Archaeology of the 1960s? Never happened. The feminist critique in the 1970s and 1980s or the reaction to the "scientific archaeology" known as postprocessualism in the 1990s? The three-winged bird had never bitten.

What this strange place had produced was some very talented theoreticians, Mayanists, and graduate students. Professors Rosemary Joyce and Bob Preucel were among some of archaeologists who were able to see the transformations taking place within the field. Later, Bill Fash was greatly influential in working with Native American students at Harvard and descendant communities at Copan in Honduras. Because of their influence, Harvard has probably produced more Native American archeology PhDs than any other school in the country. I did not know it at the time, but I had entered the institution at precisely the right moment. Harvard had renewed its commitment as one of the original Indian colleges. In 1999, the university acknowledged this position in its 1638 charter with a plaque in Harvard Yard. Unbeknownst to me, I was to be a part of a unique class of about five of the first Native American archaeology PhDs in Harvard's history.

I was introduced to the people of Cochiti and the project that helped generate this book through Professor Preucel (now at The University of Pennsylvania). In 1994, Preucel had become interested in working with the Pueblo on a joint research venture. Since that time, Preucel's work has helped generate a series of symposia and publications dealing with the archaeology of the Pueblo Revolt (2000a, 2000b, 2005a, 2005b, 2006) and opened up an important dialogue between the people of Cochiti and archaeologists.

Established in 1996 by Preucel, as a collaborative research project between The University of Pennsylvania and Cochiti Pueblo, the Kotyiti Research Project was designed to serve the needs of archaeologists, historians, and the people of Cochiti pueblo (Preucel 2000, 2002, 2005a, 2005b). The primary goals of the project were both practical

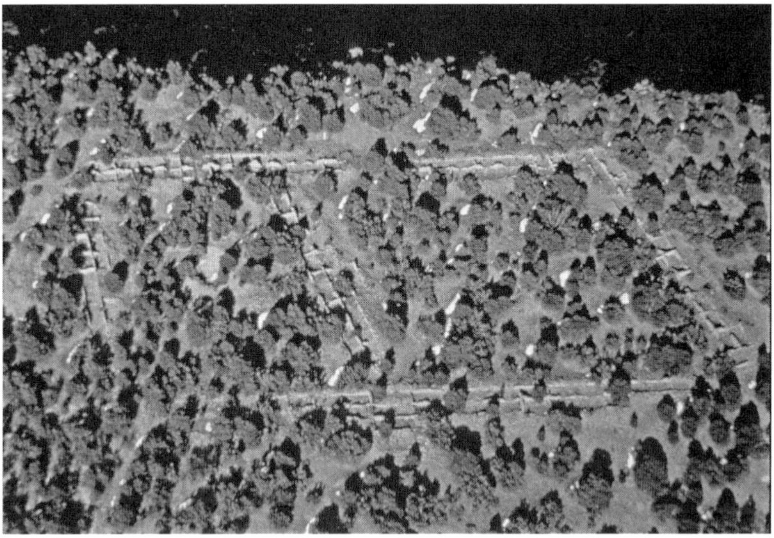

FIGURE 27. Aerial Photo of Old Cochiti (Preucel 2000).

and theoretical. Because of the centrality of Old Cochiti and the Cochiti people in the planning, execution, and defense of the Pueblos during the Revolt of 1680, an archaeological study of the processes of community formation and village construction allows archaeologists an opportunity to explore the responses and perspectives of Puebloan people to Spanish colonization. While mission settlements in the Galisteo Basin (Nelson 1914), at Hawikuh (Hodge 1918; Smith et al. 1966), Pecos (Kidder 1924, 1958), Awatovi (Montgomery et al. 1949), Giusewa (Reiter 1938), and Gran Quivira (Hayes et al. 1981) have been extensively excavated, none of Revolt period pueblos (sometimes referred to as refugee pueblos) in the Upper Rio Grande have been excavated, mapped, or analyzed with contemporary contextual archaeological methods (figure 27).

The research interests of the Pueblo were of primary concern in the project. An accurate, detailed mapping project had not been attempted since the site was excavated in the early twentieth century by Nels Nelson. Relations between Nelson and the Pueblo had been less than adequate by post-NAGPRA standards. Conducted in an attempt to secure "museum-quality specimens," Nelson's analysis was limited to excavation notes held by the American Museum of Natural History (Preucel et al. 1998: 25). Following the excavation, communication and consultation between the museum and the Pueblo were largely nonexistent. Named in oral histories, Old Cochiti is considered the home of the ancestors of the Cochiti people and continues to serve as an active component of Cochiti's religious life and social history.

Still, illegal visitations, vandalism, and pot hunting have been continual problems for the pueblo. During the project, interns were trained to assist Forest Service personnel in monitoring the site, protecting against illegal excavations and vandalism as well as stabilizing the remaining archaeological features. Working with the pueblo's language program, an oral history project was initiated to help facilitate discussions between younger members and Cochiti elders regarding Cochiti's history. The project also assisted cultural preservation officers through the collection and dissemination of the vast archaeological, ethnographic, and historic documents relating to the people of Cochiti in general and the Revolt in particular. Through a spirit of respectful dialogue and collaboration, these efforts represent an approach to archaeological research that speaks to the often unfortunate lack of communication between archaeologists and Indigenous peoples.

Among the more pressing interests of the pueblo were the exact location and jurisdiction of LA 295. Previous mapping projects had failed to determine the exact location of the pueblo. Located on Cochiti Mesa in north-central New Mexico, the community straddles the borders of Forest Service and University of New Mexico lands. Old Cochiti is known by several names; identified in Keresan as Hanut Kot-yi-ti (Cochiti Above), Hanut Katruque (Homes atop the Mesa), Kot-yi-ti shoma (Old Cochiti), K'otyiti harctitc (Cochiti Village), and Kot-yi-ti kamatse shoma (Old Cochiti Settlement), the village is well known in ethnohistoric and archaeological literature. Because of the inadequate recording and collection methods characteristic of early excavations in the region and the limited time and resources available to Forest Service archaeologists, several questions remain unexplored regarding the history and significance of the site:

- What are the relationships between the two components (LA 295 and LA 84) of the settlement?
- Given the social turmoil surrounding the rebellion, how did the people of Cochiti respond to the forced relocation, displacement, and integration of "new" community members during the Revolt? If the people of Cochiti accommodated formerly segregated members from other villages and ethnic groups, is there evidence of spatial segregation and/or marginalization?

- Is there evidence supporting Po'pay's call for a return to orthodox practices following the expulsion of Spaniards; and if so, what roles did architecture and material culture play in the pan-ethnic movement?
- Given the earlier discussion of ethnogenesis and ethnic conflicts, how can these phenomena be adequately addressed using archaeological methods?

Site Description

Old Cochiti is located at the summit of a high mesa (1,970–1,979 meters), seven miles northwest of the modern pueblo of Cochiti (see figure 3 in chapter 1). The main settlement consists of two villages, LA 84 and LA 295, several trails leading to the fertile canyons south and north of the mesa, and fortifications along the higher reaches of the trails and mesa summit. Like the majority of Revolt settlements of the upper Rio Grande Valley and Jemez Mountains, the dramatic relief between the summit and the valley below facilitated both surveillance of approaching parties and the defense of the villagers.

Described first by Don Diego de Vargas (Kessell, Hendricks, and Dodge 1995: 410), the village is referred to by the Spanish by several names. Presumably descriptions of "La Cieneguilla de Cochiti" (the Pueblo of the Little Marsh of Cochiti) and "La Mesa de la Cieneguilla de Cochiti" (the Mesa of the Little Marsh at Cochiti) refer to the wetlands that were located in the valley of the Rio Chiquito on the northern edge of the mesa. Bandelier's personal journals record the mesa and village by other names as well; "Pueblo Viejo," "Potrero Viejo," and "Potrero de las Casas" are synonymous with the mesa and the site.

The main village is an extensive rectangular construction consisting of 137 ground-floor rooms located in six semicontiguous roomblocks, framing two large plazas. Each of the plazas contains a single, large roofed kiva. Construction materials consist of shaped tuff blocks and mortar, plastered walls are still visible, and many of the standing walls reach two to three meters in height. The ground-floor rooms supported second stories in several of the roomblocks. Viga holes appear midway through several of the better preserved standing walls. Plaster is still visible along the lower and upper portions and intersections of the walls, and many of these show evidence of having been blackened by fires set by the Spanish after the pueblo was vacated in the years following the

Revolt. Seven discrete rooms appear around the perimeter of the site, and refuse middens are present along the northern, eastern, and southern edges of the main village.

The second component of the village lies 150 meters to the east of the main village. Known as LA 84, this *rancheria* style settlement consists of twenty-five discontinuous rooms. Irregularly spaced and positioned along north-south and northeast-southeast axes, the secondary village is separated by a gradual geologic saddle midway between the settlements and thick pinyon and juniper vegetation. Building materials are identical to those of the main settlement, but there are no discernible plaza areas or evidence for ritual architecture (kivas) commonly found in many other archaeological sites in the region. Two large (ten- and fifteen-meter) mounds are found at the center of the site and are separated by a long rectangular depression. Compared to LA 295, artifact density is relatively low, and a small midden area is visible along the northern exposure of the mesa's edge.

Four trails lead from the valley to the summit of the mesa. One ascends from the northeast where the Rio Chiquito empties into the gently sloping river valley. To the southwest, a second approach is possible from the higher elevations of the adjoining western mesa. Led by an auxiliary from San Felipe named Bartolomé de Ojeda in one of the sieges of the pueblo, this second trail apparently supported both livestock and horses captured from Spanish settlements. Two other, more difficult approaches are possible running up the southern and southeastern faces of the mesa. Fortifications are present throughout many points along these trails; piles of stones used to bombard the approaching attackers are still easily visible from ascending and descending viewpoints.

Cochiti Oral History: Migrations and Pueblo-Spanish Relations

The significance of the pueblo is revealed in oral histories and written records of the events surrounding the Revolt of 1680. One of Charles Lummis's Cochiti informants identified the settlement as the sixth of seven villages occupied by the ancestors of contemporary Cochiti (Lummis 1952). Perhaps the most famous of these is located within the borders of Bandelier National Monument to the northeast. Known as Tyu'non-yi 'kaih'ja (in Frijoles Canyon at Bandelier National Monument), the reconstructed Tuyonyi is one of the region's most heavily visited archaeological sites. Situated in the Cañada de Cochiti Grant (deeded to the University of New Mexico) is a third village documented in the migrational history of the Cochiti. Known as Kuapa (LA 3443, LA 3444) and regarded as a predecessor of present-day Cochiti, this expansive settlement is littered with late Kotyiti glazewares (E–F) and was most likely occupied until the late 1600s. Located near the present-day pueblo of Cochiti, Kuapa has not been mapped since the late 1800s.

Both Bandelier and Elsie Clews Parsons recorded accounts suggesting that either Tuyonyi or Kuapa were ancestral villages inhabited by the ancestors of Keres-speaking communities at Cochiti, Santo Domingo, and San Felipe (Lange and Riley 1975: 70, 398–399; Parsons 1939: 901). Ruth Benedict's (1931) collection of oral histories from the Cochiti are equally revealing. These accounts provide some contextual information supporting Puebloan notions of social distance, boundary maintenance, and the respective origins of Puebloan and European peoples.

According to her informants, the all people are believed to have emerged from Shí-pap, an underworld inhabited by animals, plants, and the supernatural companions of the Keres and other Pueblos, the katchinas. Ethnographic information does not indicate a precise location for Shí-pap, but a northerly origin of eastern Puebloan migrations is fairly common (Parsons 1939: 210–218). From here, Iyatiku, the "Mother of All," led humans from the sacred access point at Shí-pap to a village that figures prominently in Keresan religion: Kashikatchrutiya, or "White House." It is at White House that several other deities appear, and it is through their intercession that the Pueblos come to define their relationships to other kinds of peoples. From White House, Iyatiku returns to the underworld, but not before providing the Pueblos with an intermediary who aids in their survival. Appointing Tiamunyi as her representative, Tiamunyi gives the Pueblos corn and establishes a community from which all people are believed to have originated: White House.

Two other gods figure prominently in the White House pantheon. Two sisters named Utctsity and Naotsity are depicted alternately as the mothers of the Pueblos and Europeans (White 1964: 85; Parsons 1939: 244; Dumarest 1919: 212) or Pueblos and Navajos (Benedict 1931: 1), respectively.[1] The presence of these deities is significant on several levels. In one sense, Cochiti origin stories allow for the integration of Puebloan and non-Puebloan peoples within a single Pueblo-centric cosmological worldview (Preucel 1999). Utctsity is depicted as having imbued the Pueblos with virtues deemed valuable to the survival of the

Pueblos. Conversely, Naotsity is depicted as quarrelsome and arrogant.[2] Despite the nature of the contests (in which Naotsity's superior height renders advantages), Utctsity outwits her sister with the aid of the female mother figure Iyatiku.[3]

Within the setting of the contests, Utctsity affirms the superiority of the Pueblos over the "other" that opposes them. Following a series of battles in which Puebloan and European weapons are used by their respective mothers, Naotsity is killed by her sister and retreats to "Heaven where she will watch over her people. Utctsity returns to Shí-pap whence she will help the Indians with rain and crops" (Parsons 1939: 245). In the contest, Naotsity's weapons of cannon, bullet, and lance are ineffective against the Puebloan bow, war club, and spear. Naotsity "the Unhappy One" retreats, taking with her "one boy to the east, together with the images of the domestic animals and the seeds of wheat and vegetables, none of which her sister wants" (Benedict 1939: 246). The brief reference to European cultigens, weaponry, and aggression is particularly revealing. In Cochiti cosmology, the presence and rejection of Naotsity's inventory of goods affirm the location and use of "appropriate" foods, technologies, and implements for the Pueblos. Social distances are signified and affirmed through the utilization and acceptance of Puebloan material culture. More significantly, the afterlife is segregated as well; upon death, the Pueblos return to Shí-pap, while Europeans return to their mother in the European "Heaven."

White House also is the setting in which the hero twins or war gods Masewi and Oyoyewi emerge. Tiamunyi is believed to be personified by the Cacique, or spiritual leader of the Pueblo. Masewi and Oyoyewi are similarly represented by the war chiefs of the Keres Pueblos (White 1940: 96). Finally, the katchinas likewise originate at White House. Personified by Puebloan dancers during ceremonies, the katchinas embody both supernatural deities who bring rain, health, and prosperity to the Pueblos as well as the spirits of departed ancestors (Dumarest 1919: 172–183).

Oral histories also make reference to the appearance of Europeans in the Puebloan world in addition to the battles between the Spaniards and Pueblos at Old Cochiti (Benedict 1931: 187–189). Included in these accounts are references to Montezuma, chief of the Taska'la and Azteka (Tlascalan and Aztec) as well as Malinche (La Malinche) and Nan'kortez (Hernán Cortés).[4] In these accounts, most likely brought north via either traders or Native peoples left behind during the entradas, Montezuma foretells the coming of Europeans "from the northeast" (Benedict 1931: 191). The events surrounding the Revolt at Old Cochiti are remarkably similar as well:

> The Spaniards came against them again from the south, with all of the Pueblos. They laid siege to the mesa. The people went up and down through a secret trail to the north for their water. There was no path; they knew the way down the face of the mesa where there were nothing but rocks. . . . They captured a Cochiti man who had been living for a long time in Jemez . . . and made him show them the trail. . . . That night he waited at the Moon Trail and took the Spaniards and their men of the southern Pueblos up the face of the rock. . . . In the morning, the people from the pueblo woke on the mesa and saw that the whole north side of the mesa was filled with their enemies. (Benedict 1931: 186).

References to the mesa and village at Old Cochiti secure the relationship between the Cochiti people and the site, but they also leave unanswered questions surrounding the current occupational history of the Pueblo. Adding to the complexity of precontact settlement pattern history, at the current site of the Cochiti pueblo, surface collections contain sherds dating from the present through Early Glaze I and Yellow Wares (ca. 1050–1250 AD). This would indicate an occupation reaching back seven hundred years—long before the establishment of the Main Village at LA 295 (Stubbs 1950: 63, 121–122). Although constructed by Cochiti ancestors and largely abandoned as a permanent settlement following the Vargas reconquista, Old Cochiti continued to serve as a staging point for Tewa and Navajo raids in the eighteenth century.

As evidenced by historical documents and oral history accounts, Old Cochiti was an important meeting place for the leaders and planners of the rebellion. As a central community in the pan-Puebloan movement, the material culture, architecture, and site layout reflect the complexities and difficulties involved in sustaining the larger social movement that facilitated the regional rebellion. Singled out as a symbol of Puebloan resistance by Spanish colonists, Old Cochiti was unsuccessfully attacked by Juan Domínguez de Mendoza in the months following the Revolt. The December 14, 1681, attack was followed thirteen years later by a similar assault in the "bloodless" reconquest of New Mexico by Don Diego de Vargas on April 17, 1694. Sensing the dissolution of the Puebloan alliances, Vargas's attack was the first in a series of difficult and pivotal battles for control of the region. Both Black Mesa and Guadalupe Mesa were similarly targeted in battles that were socially and personally costly to both the Pueblos and their conquerors.

The ethnological significance of the community is noteworthy as well. Adolph Bandelier's sojourn among the Cochiti in the late 1800s was initiated following his correspondence with Lewis Henry Morgan (White 1940; Lange and Riley 1966: 3–20). Morgan's interest in the classification of "primitive," intermediate, and state-level societies was of interest to Bandelier, and in the winter of 1880 he began a two-month stay at Cochiti in an effort to learn the history, ethnology, and material culture of the eastern Pueblos. During subsequent years, assisted by a Cochiti informant named Juan José Montoya, Bandelier visited and recorded a number of ancestral Cochiti villages; included in his visits were Tuyonyi, Kuapa, and, most notably, "Old Cochiti," or Kotyiti (Bandelier 1880: 204). Bandelier produced two known maps of the site, the second of which was sent to the Vatican (Burrus 1969: Plate XXI).

While interdisciplinary studies of Indigenous communities are only recently returning to popularity, Bandelier's research of the history, prehistory, and ethnology of the Pueblos was groundbreaking in its scope and depth of analysis. Unfortunately, Bandelier's work took place nearly sixty years prior to the development of ceramic typologies for the Upper Rio Grande Pueblos, and his work could be characterized as "speculative" by contemporary standards. Even without formal training in ethnology or archaeology, Bandelier was able to gather and disseminate his findings to a larger audience of scholars.

Bandelier's association with the American Museum of Natural History in New York City (1903–1906) sparked interest in the site by then curator Clark Wissler. Wissler suggested that Nels Nelson explore the site, and during two weeks in the summer of 1912, Nelson excavated all 135 ground-level rooms (Preucel et al. 1998: 22). Collecting only 125 "museum-quality" artifacts, Nelson was disappointed with the effort and, more remarkably, failed to either publish a report from the excavation or create an exhibit at the museum. Aside from the two sketch maps excerpted from his notes, published information regarding the site and its significance was limited to a series of brief monitoring reports created by the U.S. Forest Service's Archaeology Division sixty-seven years later (Dougherty 1979).

Still, the excavation of a historic, nonmission Pueblo settlement represents one of the few attempts by archaeologists to chronicle the material culture and history of a recently abandoned Upper Rio Grande pueblo. Spurred by interest in Bandelier's work, ethnological interest in the people of Cochiti continued. Franz Boaz and Ruth Benedict visited the pueblo in 1924, leading to Benedict's still-referenced classic account of Cochiti oral history, *Tales of the Cochiti Indians* (1931).

Historical Background: Old Cochiti in the Colonial Spanish Documents

References to early interactions between Spanish colonists and the people of Cochiti are surpassingly sparse (Lange 1990: 1–10). Because of the location of the missionary Visita at Santo Domingo, the Cochitis were not subject to the same level of constant scrutiny as their Keres neighbors. Benavides (Hodge 1945: 20) notes the presence of three conventos among the Keres pueblos. Because of the close association between Santo Domingo and the Cochitis, it is likely that Cochiti was one of these pueblos.

Old Cochiti was the scene of several failed attempts at Spanish reconquest in the years following the Revolt. Otermín's *maestro de campo*, Juan Domínguez de Mendoza, was sent to Old Cochiti to gather information about the location and activities of the Pueblos (Hackett and Shelby 1942: 237). Finding the main pueblo abandoned, Mendoza learned that representatives from all of the Pueblos (except Zuni and Hopi) had convened for meetings at the summit of Cochiti Mesa. By mid-December 1681, Mendoza decided to negotiate peace with the Cochitis on the mesa. Sending small parties of men from his camp at the main pueblo, Mendoza noted a rapidly accumulating presence of warriors on the mesa from the pueblos of Taos, Picuris, Pecos, Jemez, and Acoma; other Keres, Tewa, and Tano peoples were present as well (Hackett and Shelby 1942: 261). Recognizing the tenuous position of his party, Mendoza met with two of Po'pay's confederates, Alonso Catiti and El Ollita.

Interviews with Pueblos loyal to the Spanish suggest that by the time of Mendoza's visit, Po'pay's movement had begun to fracture. After meeting with Mendoza, both Ollita and Catiti were reported to have initially agreed to return to their villages (Hackett and Shelby 1942: 241). Apparently, the younger men disagreed with the two leaders and decided to stage a preemptive attack on Mendoza's party. Learning of the Pueblos' plans, Mendoza retreated south to Otermín's camp, where he informed the governor of his failure. Although Otermín wanted to return to Old Cochiti, he was unable to convince his followers that victory was possible.

BANDELIER (1880)

While Bandelier was able to make use of primary historical documents held in Spanish colonial archives (Lange and Riley 1966: 8–25), the full publication of the Coronado series would not occur until the 1940s. Later generations of scholars (Nelson included) would come to rely heavily on Bandelier's interpretations of both his informants' statements as well as his sometimes questionable translations of primary historical documents.[5] With access to these texts restricted to a handful of scholars, Bandelier's interpretations of the Revolt period occupational histories were limited in their usefulness. Still, Bandelier was able to collect oral histories relating to the pueblo. His journals provide an early account of the history of the village:

> The village of Kuapa was once attacked by the Tehuas and captured. The survivors retreated to Potrero Viejo; the Tehuas pursued, but their attack upon the lofty cliff signally *[sic]* failed. They were defeated and driven back across the Rio Grande, many of whom are said to have perished in the river, and the Tehuas never troubled the Keres again. In consequence of these hostilities, the survivors established themselves on the potrero for a short time, whence they descended to settle where Cochiti stands today. . . . It seems probable that the branches of the Queres now constituting the tribes of Cochiti and San Felipe once formed one group at Kuapa, that some hostile invasion caused their dispersion, one branch retiring to the south, while the other took refuge on the Potrero Viejo and built a temporary village at least on top of this almost impregnable rock. I regard it as not at all unlikely that the aggressors were the Tehuas, since this has been told me by the people of Cochiti on many occasions. (Bandelier 1892: 165–166)

From the textual evidence provided by Vargas's personal journals, Bandelier corroborated accounts of the siege and its aftermath from his Cochiti informants (Bandelier 1892: 177–178). His description of the village is similarly informative:

> It [LA 295] is two stories high in some places, very well preserved and built of fairly regular [blocks] of tufa. The woodwork in it was evidently destroyed by fire, and much charred corn is found in the ruins. The average size of the 118 rooms on the ground floor, which are all in the pueblo with the exception of about ten, is 5.0 by 2.8 meters (16 ft. 5 in. by 9 ft. 2 in.). This is a large area in comparison with the size of older ruins. I noticed but one estufa and the pottery bears a recent character. (Bandelier 1892: 177–178)

From his measurements, he was able to produce a sketch map and a watercolor print. Still, his interpretations of the occupation and abandonment dates were less useful. Holding to his interpretations of the Vargas journals, Bandelier argued that the village had not been built until after 1687, when the Revolt had already occurred (Bandelier 1892: 167, n. 3).

Bandelier also held to the belief that the two villages on the mesa (LA 295 and LA 84) were the product of separate occupations. He characterized LA 84 as "very similar to the outhouses at Pecos, but utterly shapeless," and his informant suggested that they may have been storehouses (Lange and Riley 1966: 140). In his *Final Report*, he describes LA 84 as "traces of older ruins. . . . Possibly these smaller houses are the traces of the first occupation of Potrero Viejo by the Keres" (Bandelier 1892: 168). In his summary, he neglected to mention the similar characteristics of the ceramics found at both components of the village. Interpreting the scattered, disorganized nature of the secondary settlement as evidence supporting an earlier occupation, Bandelier summarized his interpretation of the two sites as follows: "The oldest ruins on the mesa, which hardly attract any attention, are those of a prehistoric Queres pueblo; the striking well preserved ones are those of a village built after the year 1683, and abandoned in April 1694" (Bandelier 1892: 178).

In summary, Bandelier's work accurately identified Old Cochiti, and he was the first Western scholar to record the significance of the village. His descriptive interpretations were useful in establishing the authenticity of the site; but without an accurate ceramic sequence, he was unable to determine either the construction dates or the relationships between the two villages.

NELS NELSON (1914)

Nels Nelson was the second archaeologist to work at Old Cochiti. Intrigued by the historical references provided by Bandelier, Clark Wissler suggested that he initiate a long-term research project among the Upper Rio Grande Pueblos. Correspondences between Nelson and Wissler, published by Preucel (Preucel et al. 1998: 23), provide a more complete picture of the scope and objectives of the Nelson's excavation:

> There is near Cochiti a site known as Potrero Viejo occupied from 1683–1694. This site is, I believe, just north of Cochiti and has been fully described by Bandelier. As the pueblo was burned by the Spanish, most of the objects must still be in place. I should advise careful examination of this site as it appeals to me as one of the most promising for historical problems. In this case we know

who lived there, when they came and why they left, giving us a definite section of their culture. (Preucel et al. 1996: 23)[6]

Hopeful of deciphering the relationships between abandoned communities and their descendants, and directed to collect whole pots for exhibition, Nelson began his field season in the abandoned region ancestral to the Piros (or southern Tewa) in the Galisteo Basin. After three frustrating months working at the site of San Cristobal, Nelson turned his attention to Old Cochiti. Working from inaccurate descriptions provided by Bandelier's *Final Report* and after some difficulty, Nelson was able to locate the site. Finding Bandelier's descriptions and maps of the site difficult to follow and inaccurate, he conscripted "Mexican" laborers from the Pueblo of Cochiti to work on the project. In just two weeks, Nelson had excavated the entire main village, including one previously undiscovered kiva found in the plaza area.

Counting 145 rooms, Nelson found evidence lacking supporting a multistoried habitation. His cursory analysis of the pottery fragments at the site corresponded roughly with those he had encountered in the Piro region, but he was both disappointed with the quality of the whole pots he found and unable to develop any stratigraphic sequence for the region. His excavation notes have proven to be of little value in determining more than the location of the "specimens" he collected. Although he excavated a single room at LA 84, he left no record of his findings and offered scant interpretation of the relationships between the two communities. Still, his description of the artifacts recovered leave little doubt about the role of Spanish material culture at the site. He describes finding

nine metates, some twelve or fourteen mullers, one large bowl, one small bowl, three miniature painted jars and five miniature bowls, several rubbing or hammer stones, quite a few floor polishing stones, two or three stone axes, six large, very crude stone hammers—probably used for shaping stones, one cooking slab (broken) two fire stones, a few quartz crystals, some red and yellow ochre, two or three mortar like specimens, one single bone implement, several scraps of copper plate, fragment of corroded iron, parts of a broken censer, and a portion of a small implement of copper. Besides these artifacts, we packed a few ears of charred corn, a few animal bones, a fairly complete lot of potsherds—among which may be the fragments of two or more entire vessels. (Preucel et al. 1996: 25)[7]

Nelson's assessment of LA 84 appears to have been influenced by Bandelier's *Final Report*. Possibly pressed for time, Nelson excavated a single, unnamed room and assumed that the secondary settlement resulted from an earlier occupation,

> About 500 feet southeast of this well preserved historic site [LA 295], i.e., approximately midway on the potrero, there are also faint traces of a number of ruined small houses and possibly a kiva of more ancient date. The builders of this pueblo are said to have been ancestors of the Queres Indians; but when they occupied the mesa is uncertain. The apparent age of these ruins, as judged from their advanced state of reduction, may however have some other explanation. The building stone found here might have been used in the construction of Pueblo Kotyiti. (Preucel et al. 1996: 29–30)

Without a clear sense of historic pottery types, Nelson was unable to correlate a general occupational sequence for the secondary settlement. Like Bandelier, he interpreted the irregular and complicated site plan as evidence for a much earlier occupation. While his informants (unnamed) seem to have indicated that the site was of Keres origin, his assessment of the secondary village is ambiguous, to say the least.[8]

Frustrated by the lack of whole pots and artifacts, Nelson returned to the museum hoping to create a diorama of the site. Ultimately, the museum was unable to create the exhibit he envisioned. Aside from his brief field notes, no further work was attempted at the pueblo until the Forest Service monitoring reports of the 1970s (Preucel et al. 1996: 25).

KOTYITI RESEARCH PROJECT (1996–PRESENT)

One of the goals of the Kotyiti Research Project was the integration of information not included in earlier investigations. To clarify the relationships between the two settlements and improve our understanding of how multiethnic settlements are manifested in the archaeological record, we integrated information from the historic period ceramic sequence (Mera 1933, 1934, 1939, 1940a, 1940b; Warren 1976, 1979). As our research questions focused on the dating of the main village as well as the contemporaneity and degree of social integration within and between the two components of the village, we initiated a mapping project and architectural analysis of both LA 295 and LA 84 (figure 28).

Using historical documents and primary ethnographic accounts, we incorporated information not utilized in previous studies. Among these were the much more refined ceramic sequence created by Mera as well as the dendrochronological samples collected from the site by William S. Stallings in the 1930s (1933, 1937). The following section discusses the

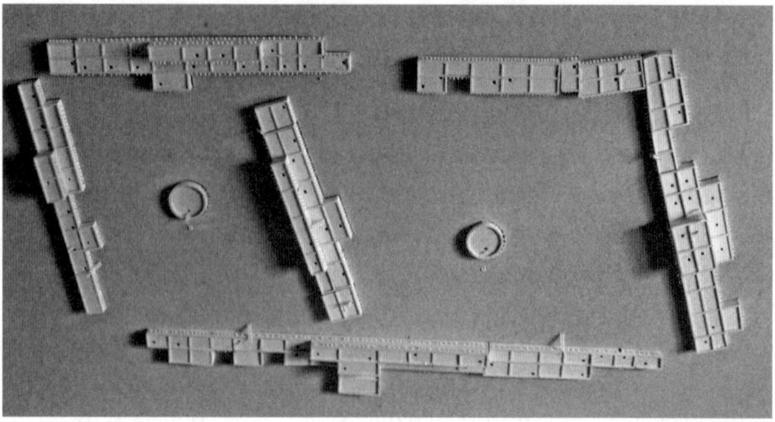

FIGURE 28. Kotyiti Research Project Reconstruction.

various components of the project. First, I explain the methodologies used to interpret the site formation processes at the two villages. Later sections describe the archaeological data collected from the mapping and architectural project at the both LA 295 and LA 84.

Architectural Information and Analysis: LA 295 and LA 84

The reconstruction of site formation processes at the main village (LA 295) required the application of a more detailed analysis of architectural features. Archaeological methods are important tools used in the reconstruction of population sizes, room functions, the sequence of construction, and the evidence surrounding the eventual destruction of the pueblo. The historical context in which the settlement was built raises a number of significant research questions and opportunities for archaeologists interested in contact period studies. Of crucial importance is the social context in which the village was first imagined and how closely this initial plan was followed as the village incorporated new individuals from formerly segregated communities. The layout of the main village, construction materials, and the use of public plaza areas and ritual architecture can provide insights into the formation, use, occupational history, and abandonment of this exceptional Revolt period community.

Roomblocks II, IV, V, and VI are separated by small passageways, characterized by Preucel (2000) as "gateways." It has been suggested that these individual passageways act as cosmological reference points

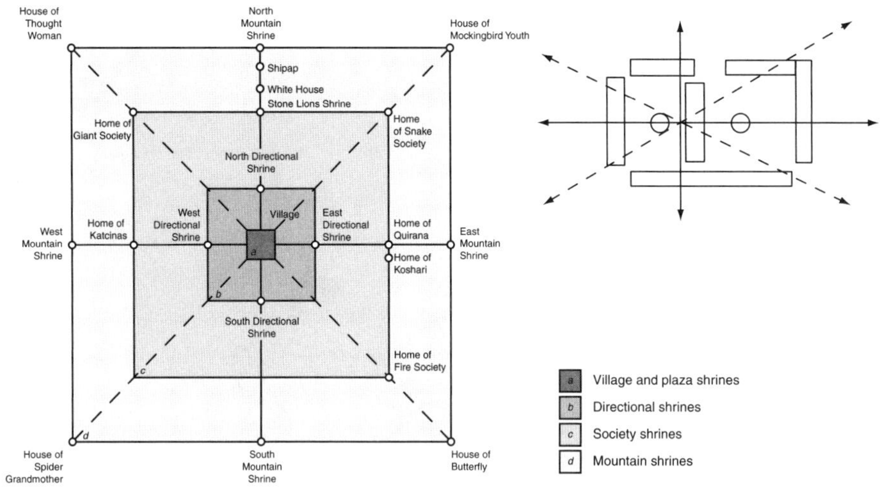

FIGURE 29. Keres Cosmological Landscape.

and situate the pueblo as the center of an ancestral Keres spiritual landscape (see figure 29). The pueblo is segregated into two discrete plaza areas, each containing medium-sized (ten-meter diameter) kivas. Roomblock VI is separated from the northern and southern roomblocks by small passageways, allowing for the free movement of people between the two plazas.

A total of 137 ground floor–level rooms were identified along with seven isolated rooms and a small shrine in the western plaza. Based on the architectural data, several building episodes were identified. The specific chronology of these episodes is unknown, although Roomblocks II and I are joined by a small section of six abutted rooms in the northeast corner of pueblo. This construction event appears to have taken place after the layout of the village was established. Based on the presence of wall joins connecting the rooms of the individual roomblocks, the village appears to have been constructed with a high degree of preparation and planning. The architectural information gathered was at times inconclusive; many of the corners are abutted at the base but joined at higher levels (the reverse was frequently true as well).

Reevaluating Community Definitions: LA 84 and LA 295 Mapping Project

One of the goals of the project was the determination of the relationships between LA 84 and LA 295. Earlier studies suggested that LA 84 was the result of an earlier occupation. However, data collected during

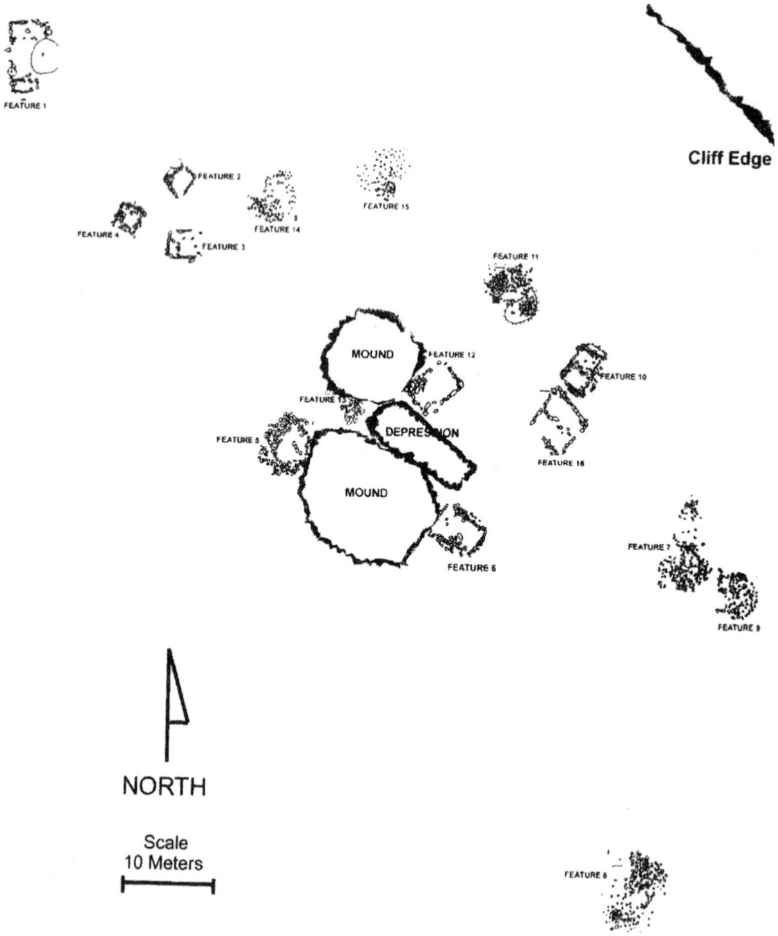

FIGURE 30. LA 84 Field Map.

the 1996–1997 field seasons suggested that the two villages were occupied at the same time. Based on the similarity of ceramics between the two sites (summarized later), it is clear that the two settlements were contemporaneous. Before determining the function of LA 84, it was necessary to develop a more accurate map of the secondary settlement (figure 30).

LA 84 is an irregular arrangement of approximately twenty-five separate features located in a thick stand of pinyon and juniper bushes. The vegetation density is high at the site. Coupled with the irregular layout

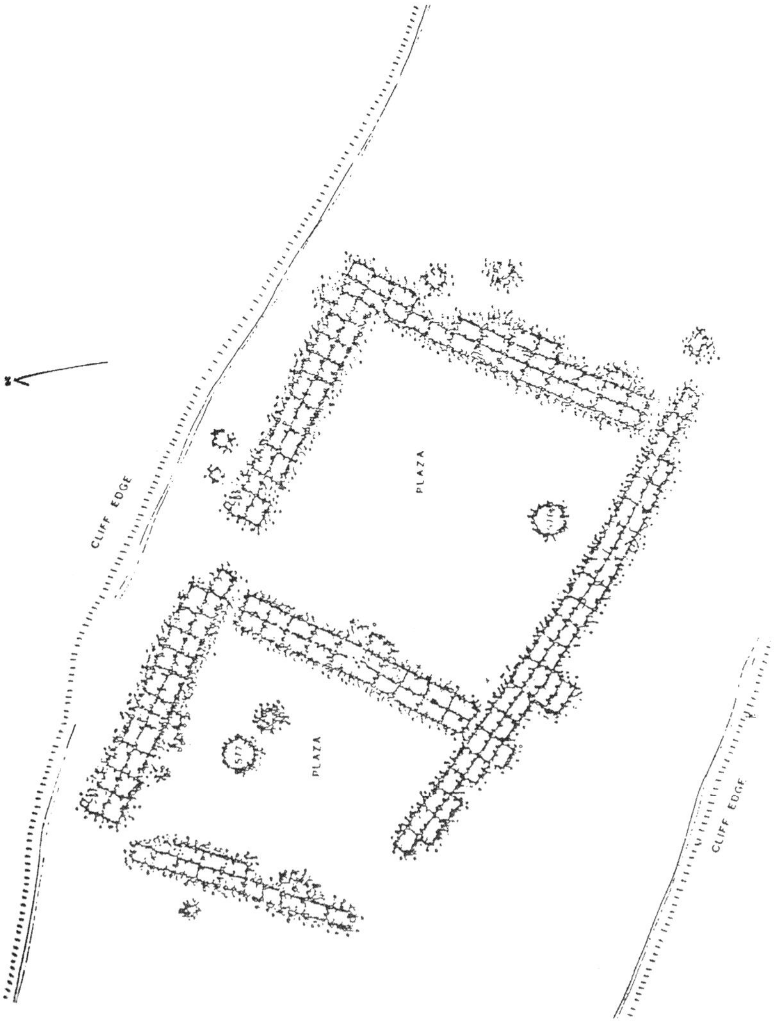

FIGURE 31. Dougherty's Map of LA 295.

of the individual rooms, this presented some challenges during the first phase of the mapping project. As evidenced by the rough sketch produced by Dougherty (figure 31), the site is difficult to orient and organize according to a more regular site layout. Because of the disjointed and scattered nature of the separate features as well as the vegetation present, LA 84 is impossible to map from a single, centrally located datum. Instead, the entire perimeter of the known features was flagged and linked in a series of fourteen points. To define the size and shape of the site, each flag linking the perimeter of the site was visible from the previous as well as following flags.

Discussion and Conclusions: Social Boundaries at the Site Level

Several conclusions can be drawn from the analysis of architecture, site layout, and artifacts present at Old Cochiti (figure 32). A careful analysis of artifacts from LA 84 suggests that both sites were occupied during the same period. More importantly, LA 84 contains artifacts consistent with permanent habitation. Domestic activities are evidenced by the presence of utility wares, storage jars, and bowls, as well as cooking and food-processing implements. The presence of hammerstones, flakes, cores, manos, and metate fragments as well as spindle whorls and shaped sherds indicates a wide range of activities. Based on this information, it is difficult to argue that the sites were used for different purposes. The ceramics present at LA 84 not only lack earlier types but almost exclusively contain glazewares of the late seventeenth century.

Po'pay's call for the abandonment of Spanish influences is supported by the artifactual evidence at Old Cochiti. Only a few Spanish artifacts were found at either site. The discovery of the censer, a device used in Catholic ceremonies, is remarkable for several reasons. There were no obvious Spanish utilitarian implements found at the village. From the presence of the censer, one can speculate about the Pueblo's appropriation of "sacred" ceremonial objects at the site. How or even if this device was used is open to interpretation, but it is clear that the possession of sacred Spanish objects would have been significant given the focus of the attack on the implements of Spanish religious ideology. The artifactual evidence raises another important question: if both sites were contemporaneous, how do we explain the association and presence on the mesa?

The answer may lie in the architecture and site plans of the two villages. At LA 295, the architectural information and construction sequence reveals a well-planned community, with a high degree of social cohesion. Built during the revivalist movement of the Revolt, LA 295 possibly reflects a return to the ancestral home of the Keres—White House (Preucel 2000). The architecture directs attention toward the center of the pueblo from a series of cosmographic reference points.

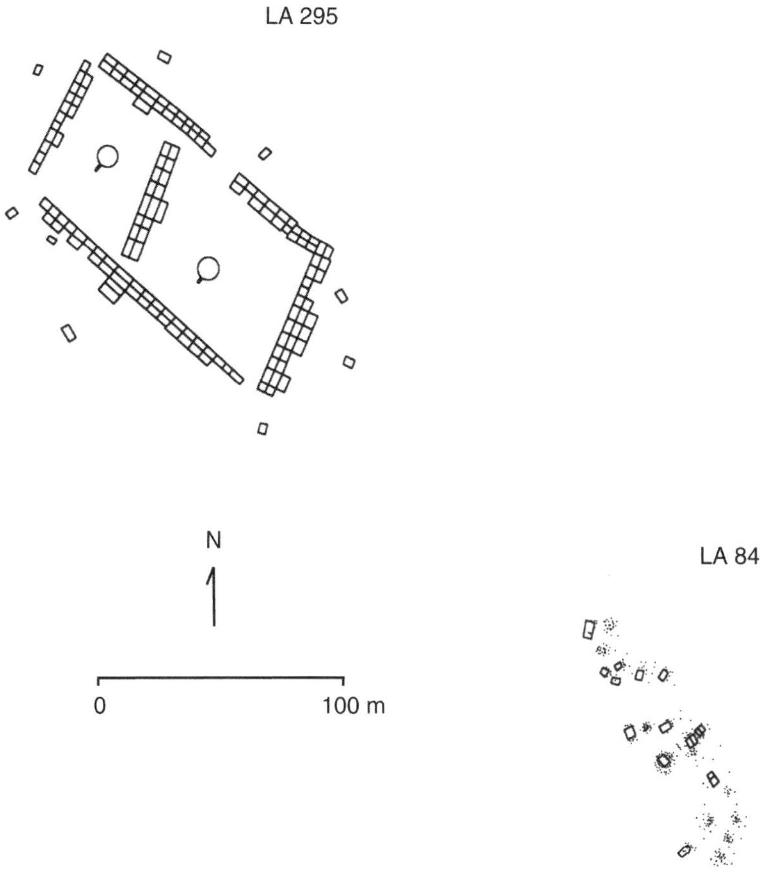

FIGURE 32. LA 84 and LA 295 Site Relationships.

Given the context in which the sites were built and occupied, the village was an important center of Pueblo resistance and ethnogenesis.

In contrast, LA 84 lacks a plaza and ceremonial architecture, and none of the walls were attached. This possibly suggests that the occupants were members of the pan-Puebloan alliance who were not accommodated ritually and socially within the main village. Segregated by 150 to 200 meters, this secondary village, I suggest, was established to maintain social distances between the communities.

The inhabitants of LA 84 may have been Apaches or Navajos, or it is also possible that the secondary village was inhabited by members of other pueblos who were excluded from the ritual activities of the main village. Other interpretations may explain the conditions and layout of LA 84. Given the fact that Po'pay was deposed during the 1680s, it is possible that the secondary settlement was composed of Keres peoples who did not completely support the objectives or behaviors of Po'pay's pan-Puebloan movement. Alternately, we may be witnessing the creation of "new" kinds of Indians such as the genízaros or Indigenous migrants, freed slaves, or traders from surrounding Indigenous communities. Preucel (2002) has argued that this village arrangement may have incorporated Tewa peoples. They may have been allowed to seek the protection of the Cochiti villagers but perhaps were not welcome to create their own ritual architecture. In either case, the integration of new narratives that seek continuity and emphasize the study of ethnicity as a process opens the door for new interpretations of site and village organization.

CHAPTER 8

Repatriating Old Cochiti

In old Santa Fe we have La Fiesta
'Tis the time for singing, dance and play
On this day we do not take la siesta
While Zozobra burns the gloom away
Steel guitars are softly strumming music
Senorita, come along with me
Luminarias all are shining brightly
At the baile, fancy costumes you will see

Si Señor, como no, let's go out and have some fun
And we'll see Santa Fe in Conquistadores way.

Song of the Fiesta of Santa Fe

The Fiesta of Santa Fe is held each fall in the Old Town area surrounding the Old Governor's Residence. The fiesta blends Catholic piety and the veneration of Mary with New Age mysticism and the lively mariachi music and food typical of Mexican fiestas. The ceremony takes place in open arenas and is decidedly secular; bar hopping and psychedelic drugs are often associated with the strange vision of a burning papier-mâché likeness of "Old Man Gloom"; this "Burning Man" effigy, or Zozobra, was invented in the 1920s by a local artist. This symbolism is jarringly juxtaposed with the celebration of the patron saint of the Chapel of Nuestra Senora de la Reconquista, Our Lady of the Reconquest. Originally the fiesta commemorated the removal of a small wooden likeness of Mary, mother of Jesus, from the Spanish church during the

Revolt of 1680. The small twenty-one-inch likeness is purported to have been a gift of Fray Alonso de Benavides (author of the Benavides memorial) to the colonists in 1625. The statue is said to have spiritually inspired and aided Don Diego de Vargas in his efforts to reconquer the Pueblos during the misnamed "bloodless reconquest" of 1692.

The irony of the statue and the ceremony commemorating the "reconquest" is not lost on the Pueblos, who refuse to participate in the ceremony or the fiesta. In recent years, Anglos, dressed in fake war paint and dime store burlap Indian headdresses, so offended local Indian people that the fiesta has remained off limits to many. The whole idea that the city might celebrate the burning of Pueblo villages and the costly battles against Pueblo ancestors demonstrates the enduring schism between the narratives and mythologies of Spaniards and Indians. The notion that the mother of Christ might be associated with violent military encounters against the Pueblos is similarly difficult to fathom for many Pueblos.

These associations are further complicated by the fact that the Pueblos are among the most devout Catholics I have ever met. Having lived with a conservative religious family in Seville, Spain, and having witnessed the pageantry of Semana Santa firsthand, I would suggest that both groups are equally fervent in their veneration of Catholicism. The difference is that these two versions of Jesus and the Church are enshrined in parallel and contradictory historical narratives. Our Lady of the Conquest has been rehabilitated as well by contemporary Catholic clergy. Similar to the manner in which jihad is internally contested as a symbol of both violent struggle against nonbelievers and an internal struggle between God's will and the will of the individual, Our Lady of the Conquest is now represented as "a conquest of the Heart," not an emblem of religiously sanctioned violence. The tensions are still there, but the story of the Pueblos and their powerful counternarrative to conquest are increasingly recognized in popular discourse (figure 33).

Former All Pueblo Council president Hermann Agoyo clearly remembers having never heard of the Pueblo Revolt until he was *out of college*—a remarkable testament to the power of colonial narratives in New Mexico. Thankfully, this perception is changing, and Native peoples are beginning to assert their narratives and dismantle the mythology of conquest with tools more powerful than the jackhammer in Santa Fe. The installation of the Po'pay statue in the Hall of Heroes in Washington, DC, is one powerful manifestation of that shift. A second, more minor victory may be claimed by the pueblo of Cochiti. In 2005,

FIGURE 33. Colonial Statue in Santa Fe Plaza.

after years of stonewalling and fruitless appeals, the University of New Mexico relented on its claim to the ancestral village of Old Cochiti. Governors Henry Suina and former Cochiti governor Joseph Suina lobbied intensively throughout the past twenty years to have the village formally repatriated. Information and documentation gathered by Robert Preucell's Kotyiti Research Project were made available to the Pueblo in the hopes that such an accommodation and recognition of Cochiti's history might occur. For me, this represented the material manifestation of a decade of work and a deeply held belief that archaeologists and Native peoples might work together in rendering a more complete and active version of the past. Imagining an Indigenous archaeology is but one manifestation of this movement. The slow dismantling of the mythology of conquest and the invisible Indian is a larger project, but some inroads toward such a goal have been made. An Indigenous archaeology, however formulated, incorporates multiple perspectives on the past and can only lead to the development of a healthier, more intellectually resilient field, conscious of but not limited by its colonial associations. I believe that collaboration with descendant communities *is* the future of archaeology in North America.

Implications Archaeological and Historical

In the beginning of this study, I suggested that for reasons both theoretical and political, the interpretations of many archaeologists and historians had contributed to a sense of discontinuity between an Indigenous past and a colonial present. Processual emphasis on disease and the reluctance of historians to engage in discussions of Spanish colonial violence had helped generate interpretations of the colonial period that failed to explain either events such as the Revolt of 1680 or the continued presence of contemporary Native Americans. Introducing the concept of Indigenous archaeology, I argued that a more fruitful analysis of contact period resistance and change was provided by an application of theories of cultural interaction collectively known as ethnicity. From here I explained how ethnic conflicts are generated and resolved through the processes of ethnic conflict.

The evidence presented in this book clearly refutes the idea that Puebloan societies were wholly transformed by disease, acculturation, conquest, and the introduction of Spanish technologies during the sixteenth and seventeenth centuries. Archaeological and historical documentation of contact and Revolt period sites reveals a very different process of social interaction, culture change, and resistance. The following conclusions can be drawn from this work.

Ethnicity and ethnic conflicts are social processes that can be identified through contextual archaeological methods.

While I have positioned part of this work as a response to some of the shortcomings of processual archaeology, I do not believe that ethnicity theory and processualism are in any way incompatible. Processual archaeology is alive and well in the Southwest, and it foolish to imagine that one book will change that. Still, theories of ethnicity have a processual dimension that can be applied in other temporal and geographic settings.

Material technologies are embedded within symbolic systems.

Narratives of cultural interaction that emphasize the adoption of "superior" European technologies as evidence supporting acculturation or disappearance of "traditional" cultures need to be reevaluated. Many of these interpretations do not consider the symbolic associations and actor-centered perspectives of these objects within Indigenous societies. Working against the tenets of behavior posited by "rational economic man," this Indigenous approach to materiality, contact, and colonization demonstrates that people and groups do not always choose to adopt more efficient or economically more productive modes of subsistence. The power relations between groups seem to determine how material culture, symbolic systems, and new technologies are incorporated (or not incorporated) by subordinated populations. This contextual study of colonial interaction provides a counterargument to many of the acculturation theories advanced in the last fifty years.

Ethnic identities, codified in specific ways of considering the self in relation to the group and the group in relation to other groups, are conservative in these situations. Despite the pervasive narrative of conquest and acculturation, power relations figure prominently in the acceptance or rejection of new technologies or ideological systems; awareness of local identities is *heightened*, not dissolved, in these circumstances. Forcible acculturation, mass baptisms, and military victories do not automatically lead to the adoption of Western modes of behavior or the dissolution of Indigenous cultures. In fact, the opposite result is often achieved.

Even when aspects of colonial cultures are adopted by Indigenous peoples, they are often articulated in ways that are intelligible only to the subjects themselves. Ceremonies and symbolic representations are often reproduced in ways that are unintelligible to the dominant culture. The public performance of these rituals are inherently ambiguous to outsiders (e.g., the corn dances on feast days) and reveal to community members the maintenance of traditions and identities that are only knowable to the members themselves. In this sense, they demonstrate the maintenance of power over symbolic systems. Catholicism was adopted, but it was articulated in ways that were very different from the meanings intended by colonial missionaries.

There are several important limitations to this kind of approach to the past, but these are themselves instructive—particularly in light of contemporary archaeology's struggle with the processes, methods, and implications of NAGPRA. There are limitations on the precise association of style and technology with Indigenous cultures; if we cannot determine cultural affiliation through a direct association of style, we need to acknowledge that these limitations may never be overcome without extensive documentation. However, this does not exclude the possibility of these relationships can and do exist. This leads to a larger question regarding the power of archaeological discourse as a vehicle through which prehistoric materials and contemporary Indigenous peoples are alienated. Change and population movements do not always signal the end or the death of Indian cultures. The mythology of conquest and the invisible Indian are largely based on this assumption.

The legacy of acculturationist models within ethnographic studies should be carefully considered. Ethnohistoric information is not an accurate proxy for the determination of dynamic and evolving meanings and significations of Indigenous cultural materials. Too frequently, ethnography is depicted as the final authority in the determination of cultural authenticity. Based on the power relations between Western ethnographers and Indigenous informants, much of what we know is fragmentary and must be acknowledged as such. Just because the Pueblos have not chosen to reveal their histories or explain the details of their religious practices does not mean that such knowledge does not exist.

The use of historical sources for population estimates and the study of colonial violence must be considered in congruence.

Following the critical engagements with historical, ethnographic, and archaeological materials proposed in Indigenous archaeology, historical sources themselves need to be considered in light of the contexts in which they were produced. Black Legend revisionism amplifies the transformative effects of military victories by Europeans but minimizes and subordinates the long-term, covert strategies of retaliation and resistance that are generated by these actions. Black Legend revisionism may have resurrected the status of Spanish colonization vis-à-vis the Anglophile narrative, but it also produced a body of work that deemphasized the use and transformative consequences of colonial violence. Indigenous archaeology is not an attempt assail Western culture or write more "polite" histories of contact; rather, this approach recognizes social stigmatization, marginalization, and subordination as processes that can be studied comparatively in other contexts. The use of legitimate, institutionalized forms of violence was (and is) not limited to Spanish or even Western cultures; but without a serious discussion of the methods of subordination, we are unable to explore the social and demographic responses of colonized peoples to subordination. The Revolt is only one such example.

Historical population estimates require a similar level of contextual scrutiny. Early estimates (influenced by a desire to demonstrate the wealth of human resources or the military prowess of Spanish conquistadores) are almost always inflated but are often uncritically deployed to provide evidence supporting large-scale populations crashes (see table 8.1 and figure 34). Without contextual information, missionary records are equally unreliable. These estimates represent the participation of Pueblos in Catholic sacraments. Given the focus of the Revolt on the rejection of Catholicism and the affirmation of traditional Puebloan religious practices, it is questionable to assume that these counts serve as a clear proxy for actual population numbers. The coincidence of population crash with the Revolts of the seventeenth centuries are highly suspect given these contexts. These were simply not scientific censuses.

Disease-based theories should be similarly reevaluated. Disease allowed historians to posit alternative (nonviolent) explanations for population crashes, but these do not adequately address the human causes and social consequences of colonization. Excessive tribute requirements and the appropriation of land, water, labor, and resources compromised the health of Native communities. The prominence of disease-based explanations and the avoidance of discussions of colonial violence lead us to make inferences about population crashes and the cultural consequences of those crashes. What the documents and archaeological data reveal are the abandonments of villages in order to resist colonists, maintain social distance, and preserve the symbolic and material resources of the Pueblos. The consequences and implications of these two events are very different. The deployment of disease as an explanation of abandonment supports the colonial trope of Terra Nullis

TABLE 8.1. Population Estimates of the Pueblos during the Entrada Period (1539–1602)

	Year of Testimony	Entrada/Chronicler	Method of Counting
A)	1539	Marcos De Niza	Nonvisual estimate
B)	1560	Coronado (1540–1541)/Pedro de Castaneda	Secondhand accounts and visual estimate
C)	ca. 1582	Chamuscado Rodriguez (1581–1582)/Hernan Gallegos	Visual estimate
D)	1583	Chamuscado Rodriguez (1581–1582)/F. Escalante, H. Barrando	Visual estimate
E)	ca. 1583	Antonio De Espejo (1582–1583)/Diego Perez de Luxan	Visual estimate
F)	1583	Antonio De Espejo (1582–1583)/Antonio de Espejo	Visual estimate
G)	1590	Castano de Sosa (1590)/Castano de Sosa	Visual estimate
H)	1598	Juan Morlete (1591)	
I)		Leyva-Humana (1593)	
J)	March 2, 1599	Don Juan de Onate (1598)/Don Juan de Onate	Visual estimate after 10 months (letter to King)
K)	March 22 1601	Don Juan de Onate	Visual estimate after 22 months
L)	July, 1601	Interview of soldier, Jusepe Brondate	Unknown
M)	July 1601	Interview of soldier, Marcelo de Espinosa	Unknown
N)	July 1601	Interview of soldier, Gines de Herrera	Unknown
O)	October 1, 1601	Letter written by Franciscan priest, Juan de Escalona to the Viceroy	Visual
P)	October 2, 1601	Petition to the Viceroy for aid by Captain Geronimo Marquez	Unknown
Q)	April 23 1602	Interview of Baltasar Martines, soldier sent to reinforce the colony in Dec. 1600	Heard estimates from those who "had traveled over this land"
R)	April 26, 1602	Testimony of Juan Rodriguez, a Portugese soldier	Visual estimate: "visited all of them"
S)	1605	Fray Francisco de Escobar, letter to Viceroy	Unknown
T)	April 9, 1609	Fray Francisco de Velasco, letter to the King	Unknown
U)	1626	Missionary/Fray Zarate Salmeron, Memorial, plea for resources (Salmeron 1966:35)	Unknown
V)	1626	Missionary/Fray Juan de Santander (Ayer 1965:6)	Unknown/written from Spain
W)	1630	Missionary/Fray Alonso de Benavides (Hammond and Rey 1945:35)	Unknown/visual estimate
X)	1638	Missionary/Juan de Prada (hackett 1937:108)	Never there
Y)	1660–1690	Missionary/Fray Agustin de Vetancurt (Bandelier 1890:122)	Total New Mexico Population (Spanish and Indian)
Z)	1678	Cuistodia of New Mexico/Fray Francisco de Ayeta (Hackett 1937:299)	Baptized
AA)	1706	Missionary/Fray Juan Alvarez (Hackett 1937: 373–377)	Christian Indians
BB)	1744	Missionary/Fray Miguel de Menchero (Hackett 1937:402–407)	Christian Indians

NOTES: C) This is a report prepared by Hernan Gallegos, a Penninsulare born in Sevilla and in New Spain for 7 years. He was a slave trader and soldier who lived in the Northern Frontier. The report was to the Viceroy of New Spain, Mendoza (Hammond and Rey 1966:102–108). Gallegos makes no attempt to estimate the population in three documents. His fellow soldiers, Escalante and Barrando state in their own testimony that, "We agreed that in sixty-one pueblos that we saw and visited, there must be more than one hundred thirty thousand persons" (Hammond and Rey 1966:142).

D) Fellow soldiers, Escalante and Barrando state in their own testimony that the leaders of the entrada collectively estimated a large population. Chamuscado died on the trip back to Santa Barbara, "We agreed that in sixty-one pueblos that we saw and visited, there must be more than one hundred thirty thousand persons" (Hammond and Rey 1966:142).

E) Luxan was a notary (chronicler) and soldier in the moning district. He mentions visiting approximately sixty pueblos. At only thirteen of these, does he mention any population at all. At thirteen of the sixty, he estimates the presence of 33,300 people (Hammond and Rey 1966:172–185).

F) Espejo wrote this as a report, presumably to the Viceroy. He was a peninsulare, born in Cordoba, Andalucia. He was a fugitive nobleman who had fled from Mexico City after being charged with murder. He was a cattleman and rancher as well as a soldier in the mining district (Hammond and Rey 1966:15–28).

J) This number is taken from Onate's first report to the Viceroy, ten months after first contacting the pueblos. He states that there are "conservative in my reckoning, sixty thousand Indians" (Hammond and Rey 1953:483).

K) This letter was written to his family in Spain after twenty two months in New Mexico. No explanation is given for the decline in his estimate (Hammond and Rey 1953:619).

L) This estimate is taken from an interview of one of the soldiers in Onate's party, Jusepe Brondate, in an investigation of Onate, ordered by the Viceroy. The investigator is Francisco Valverde (Hammond and Rey 1953:629).

M) Interview of soldier Marcelo de Espinosa in the same report (Hammond and Rey 1953:639).

N) Interview of Gines de Herrera in the same report (Hammond and Rey 1953:652).

O) This figure is taken from a letter written by one of the Franciscan priests present in Onate's expedition to the Viceroy, reporting that the populations Onate has estimated are highly inflated but that the colony should be maintained anyway. Normally, the priests' estimates are among the most inflated, but Escalona states that, "Eager to bring good tidings to his magesty, some people have given free reign to their pens, telling of things which do not exist in this land, making provinces out of puebos (they called Taos and Santo Domingo Provinces). Similarly, they describe the other pueblos as provinces, even though the largest will not contain more than two or three hundred people. The entire land discovered thus far does not contain 22,000 Indians" (Hammond and Rey 1953:695).

Unit Counted	Number of Unit Counted	Dates: Amount of Time among the Pueblos
Homes	100,000 at Zuni alone	
Total population	20,000	July 1540–March 1542 (21 months)
Appx. 71 villages	Does not mention population	August 1581–January 1582 (5 months)
Persons	130,000	
Appx. 60 villages	(33,000 at 13 of 60 villages). He does not mention populations at the others	February 1583–July 1583 (5 months)
61 villages	253,000	
	No estimate of population is given	December 1590–June 1591 (7 months)
	No estimate of population is given	March 1591–June 1591 (4 months)
	No estimate of population is given	January 1598–December 1598 (12 Months)
Indians	60,000	May 1598-Permanent colonization (estimate is given after 10 months)
Persons	(40–50,000) 45,000	22 months
Persons in 130 pueblos	(50–60,000) 55,000	26 months
Persons in 125–130 pueblos	(22–24,000) 22,500	26 months
Persons in 125–130 pueblos	30,000	26 months
Indians	(less than) 20,000	29 months
People	(more than) 60,000	29 months
People	16,000	11 months
Indians	(30,000 plus women and children)	Unknown
Souls	30,000	Unknown
Souls	30,000	Unknown
self-reported baptisms	34,650	1621–1626
"Indian population"	34,320	Never in New Mexico
Self-reported baptisms	80,000	1623–1629
Self-reported baptisms	40,000	Never Present in New Mexico
Self-reported baptisms	24,000	Present in New Mexico-exact dates unknown
Self-reported estimates	17,000	1675–1678 (In transit between New Mexico and Mexico)
Christian Indians	8,790	Unknown
Christian Indians	6,795	Unknown

P) This is a leter written by one of Onate's captains, Geronimo Marquez. It is a petition for aid and was written in response to the desertion of a large number of Onate's party due to a lack of pueblo people and natural resources. Marquez asks for aid to support new converts to Christianity (Hammond and Rey 1953:702).

Q) This is taken from the Valverde inquiry, ordered by the Viceroy to determine what kind of settlement Onate had been found. "He heard from many who had traveled all over the land that there were about sixteen thousand people settled peacefully in the territory under his jourisdiction." The soldier, Baltasar Martinez, returned from New Mexico with news that only about "thirty youngsters and twenty women slaves had been baptized . . . the friars believed that the land was poor and would not be maintained permanently." He is in favor of maintaining the colony only if the colonists are supported by the crown (Hammond and Rey 1953:838, 850).

R) This testimony was taken from a Portugese soldier assigned as a Comander in the company of Captain Zaldivar. The reason for his return is not explained (Hammond and Rey 1953:863).

S) This is copied from a letter or diary entry written to the Viceroy by Father Francisco Escobar. Escobar delivered his account of Onate's trip to California in an effort to convince the Viceroy and King not to abandon the colony. Viceroy Montesclaros, having read through all available testimony and heard Escobar's testimony on October 25th, believed that the colony was doomed to fail and that Onate and his supportes had deceived him. "From his (Onate's) signed report and from four witnesses who returned with the friar, I cannot help but inform your majesty that this conquest is becoming a fairy tale. If those who write the reports imagine that they are believed by those who read them, they are greatly mistaken. Less substance is being revealed every day" (Hammond and Rey 1953:1012,1009).

T) This is a letter written by Father Francisco Velasco to the King in April, 1609. He is pleading for the continuation of support by the crown for the colony. He claims that there are "thirty thousand souls, most of whom are clamoring for baptism and conversion . . . " (Hammond and Rey 1953:1092).

U) Father Zarate Salmeron arrived in New Mexico in 1621 to serve as a Franciscan "Comisario" from 1621–1626. He established a mission among the Jemez (San Jose in 1621) but by 1622 the Jemez were no longer under control of the Spanish. Evidence of this is provided in 1622 by Fray Martin Arvide, a missionary among the Picuris, who requested permission to resettle the Jemez. The Jemez had revolted and abandoned the missions at San Diego and San Jose. Whatever active mission that was present was burned once again at Lent 1623 during Salmeron's tenure and never reoccuppied. There is no evidence that Salmeron was living at Jemez for the six years of hius tenure (as is often reported). His "Relacion" written from Mexico City in 1629 (3 years after his departure from New Mexico), is a plea to the crown to maintain the colony and like most of his contemporaries, he uses the Crown's obligation to Indian converts, and exaggerated accounts of mineral wealth. Salmeron claimes to have baptized 6,566 Jemez Indians. This figure is

NOTES: (*continued*)
later added to the estimates of 20,000 pueblos in the region and is often treated as an actual census figure because of the non-rounded number (26,650). There are no baptismal records that verify his claim (Ayer 1965:35).

V) Fray Juan de Santander, Commissary General of the Indies (based in Spain) recorded this figure in an introductory letter to Benavides 1630 Memorial. His population estimate is contradicted in Benavides text. Benavides' figure of 500,000 is never clarified by Santander. This is a plea to King Phillip IV for resources (Ayer 1965:6).

W) Fray Alonso de Benavides wrote his "Memorial" in 1630 for King Phillip IV. The revised edition (1634) was written for Pope Urban VIII in an effort to persuade the Pope of the importance of Franciscan missionization in New Mexico. Benavides' account is full of wild exaggerations (among them 500,000 Indians in New Mexico, an inland kingdom "Quivira", a "South Sea" and in inland ocean passage and numerous mines capable of supporting the finacial expenses of the colony). Benavides' 500,000 figure is stated clearly. Whether the 80,000 figure refers only to Indians in New Mexico is simply unknown. His account was printed in Madrid in 1630 and a revised edition, "the revised Memorial" was presented to Pope Urban VIII in 1634 (Hammond and Rey 1945:35).

X) Fray Juan de Prada was the Franciscan Commisary General in Mexico. He never claims to have been in New Mexico but was stationed in San Francisco, Mexico as the Head of the Franciscan Order's missionary efforts. In a report (a petition written September 26, 1638), de Prada argues against the creation of a Bishopric in New Mexico citing insufficient tribute amounts. In an offhand comment, de Prada suggests that his 40,000 figure "for although there must have beem more than sixty thousand baptized, today these conversions are diminished to that extent on account of the very active prevalence during these last years of smallpox and the sickness which the Mexicans call Cocolitzli" (Hackett 1937:108). This offhand reference has been repeated by archaeologists and historians (Barrett 2002:65), but is not corroborated by any other account. There is simply no mention of any epidemic in the 1600's in New Mexico in any other document.

Y) Vetancurt was a missionary in New Mexico. His "Teato mexicano" was written in the 1690's after the Revolt of 1680. He estimates the total population of Spanish, Pueblo and Indian allies in 1660 at 24,000, "Pues el ano de 1660 se hizo pardon general, en que se hallaron mas de vienticuatro mil personas, chicas y grandes. indios y espanoles" (Bandelier 1890:122).

Z) Fray Francisco de Ayeta was the Procurator General of the Franciscan missions from 1675–1678. He was the last Custodia sent to New Mexico prior to the revolt and his figure (17,000) was referenced in a letter he wrote to the king pleading for soldiers and resources to defend the colony ". . . according to the registers, there are 17,000 Christian Indians, men and women . . ." (Hackett 1937:299).

AA) These figures are among the few remaining post-revolt figures and are noted in a report written by Fray Juan Alvarez to the crown. He estimates the number of Christian Indians living in the pueblo villages. This figure reflects the number of Indians participating in Catholicism in the post-revolt era. They are individual village estimates, not counts of baptisms. These were tallied by the author (Wilcox). Found in (Hackett 1937:373–377).

BB) These figures are estimates of Christian Indians living within each pueblo (tallied by the author-Wilcox). They are not taken from baptismal records (rounded numbers are used for each village) and they likely reflect low participation rates in Catholicism among the pueblos in the post-revolt period.

and inhibits the development of scholarly methods that connect contemporary Indians to the material remains of their ancestors.

The implications are clear: if diseases caused abandonments, then there is no need to look for and document historical Puebloan settlements. This supports a view of pre- and postcontact discontinuity. If abandonments are motivated by a desire to maintain social distances and actually preserve cultural differences through conservative social movements, then the dichotomy of pre- and postcontact Indigenous cultures is weakened, and the mythology of conquest—biological, religious, and cultural—is called into question.

Other questions are raised by the application of Indigenous archaeology. The ethnogenesis of genízaro peoples has never been clearly factored into population estimates. Theories of ethnicity suggest that one option for stigmatized populations is the creation or adoption of alternate social identities. Many genízaros were descendants of Apache, Navajo, or Pueblo slaves taken in raids or from unions between Puebloan women and Spanish men. The children from these unions (as well as their mothers), often marginalized by the Pueblos, came to represent a

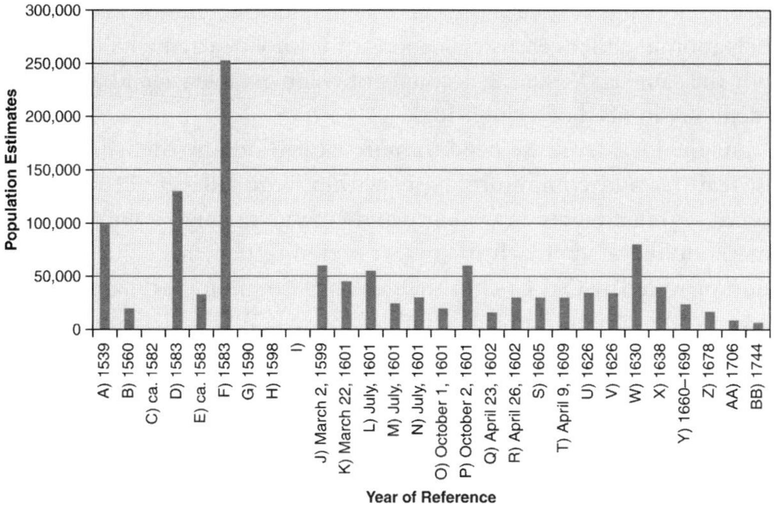

FIGURE 34. Pueblo Population Estimates.

significant portion of New Mexico's colonial population (Gutiérrez 1991: 150–156; 174–175). Schroeder estimates that by the eighteenth century, one-third of New Mexico's colonial population was genízaro (1975: 62). Forcibly relocated from Santa Fe to occupy the Borderlands of New Mexico, genízaro communities occupied a distinct, self-segregated, linguistically differentiated community of Indigenous peoples. Abiquiu, Ojo Caliente, Tomé, and Belen were genízaro communities established to absorb and segregate marginalized Indians from colonists of higher social status (Hodge 1912: 489; Jones 1966: 139).[1]

According to Gutiérrez (1991: 156), 10 percent of those persons legally married between 1700 and 1846 were *hijos de la iglesia* (children of the church). These were orphaned Native peoples. Abandoned by Puebloan people, many of these genízaros were literally laid at the doors of churches and conventos. Little is known about the role of the genízaros during the Revolt. The location, size, and loyalties exhibited by the Indios of San Miguel, the Indian village located adjacent to Santa Fe, are similarly unknown. Finally, the documentation of settlements on the northern periphery of the Jemez Mountains demonstrate the need to identify and gather more information about historical Puebloan communities throughout the region. The evidence for reoccupation suggests that not all Pueblos lived in close proximity to these missions. This raises further questions about how accurate baptismal records are.

Are we looking at total numbers of people or the rates of Pueblo participation in either census collections or Catholic sacraments? They are not the same and, like the accounts of colonial violence and abandonment, are in need of a closer look.

At the local level, we need to more clearly identify the whole range of features and community types within large villages. Processes of social segregation are in evidence and emerge through a more refined methodological approach to contact period settlements. The information provided at Old Cochiti suggests that communities may manifest social differentiation in any number of ways. While hierarchical relationships are a particular subject of interest in neo-evolutionary archaeological studies, ethnic divisions are revealed through physical distance and segregation at the site level. As the relationships between LA295 and LA84 suggest, physical distances may manifest, reflect, and reproduce social distances. The presence of these new kinds of communities requires the recognition of other kinds on Indigenous communities. To date, we have little information regarding historical genízaro (Christian pueblos) and Navajo-Apache relations (Gobernador settlements are one example). It is clear that many people defected from the pueblos and assumed social identities that would not expose them to the social costs associated with *being* Pueblo.

An Indigenous approach to archaeology must be grounded in a willingness to develop models and methods that allow for the integration of historical and ethnographic sources. The use of history and ethnography serve a different set of needs for archaeologists today. With NAGPRA comes an opportunity to grapple with new theoretical and methodological issues. Our ideas regarding cultural continuity, acculturation, and affiliation must take into account the complexities that exist in the refashioning of social and cultural identities. The dynamic nature of the processes of ethnogenesis are accelerated and heightened by colonial contacts. If abandonments reveal the "shells" of individual communities, then the job of contemporary archaeologists should include the development of explanations of Indigenous survival and continuity. Repatriation represents a new beginning and methodological commitment for archaeologists. Rather than limiting archaeological practices, the illumination of connections between contemporary Indigenous peoples and ancestral materials enriches the practice of archaeology with a new set of questions. In addressing and deconstructing the mythology of contact, conquest, and colonization, we help move the field to a new level of dialogue and engagement with Indigenous peoples. NAGPRA was only a beginning.

Notes

Chapter 1

1. "We have been taught, inside the classroom and outside of it, that there exists an entity called the West, and that one can think of this West as a society and civilization independent of and in opposition to other societies and civilizations. Many of us grew up believing that this West has a genealogy, according to which ancient Greece begat Rome, Rome begat Christian Europe, Christian Europe begat the Renaissance, the Renaissance the Enlightenment. The Enlightenment Political Democracy and the Industrial Revolution" (Wolf 1982: 5).

2. Wolf's comments were made in a lecture entitled "The People without History," delivered to the University Faculty Senate, 125th plenary session, City University of New York, December 13, 1983. The lecture is excerpted in Schneider and Rapp 1995: 6–7.

3. See Ferguson and Colwell-Chanthaphonh 2007; Hurst-Thomas 2005; Lightfoot 2005; Mills 2004; Meskell 2007; Preucel 2007.

4. See Crosby 1987 for a discussion of the "Bardic interpretation" of colonial expansion, its moral justifications, and the manner in which colonization of the "New World" was viewed as an extension of divine providence. Bancroft called for a history that explained "the steps by which a favoring Providence, calling our institutions into being, has conducted this country to its present happiness and glory" (1966: 4). Bancroft's work situated the history of the United States as a continent "uncontaminated" by what Crosby calls the "decadent" history and traditions of Europe. The counterimage was provided by the imperfect "proto-civilizations" of the Americas; having been developed by "Barbarians" and "savages," they fell quickly to the divinely chosen inheritors of the continent (1987: 1).

5. These basic definitional problems are often compounded by the failure of many scholars to properly cite or situate their work with reference to the work

of scholars outside their fields. An extensive literature deals with the complexities, inconsistencies, and usefulness of this term. These definitions are consistent with those developed in synthetic, interdisciplinary analyses offered in Cohen 1978, Glazer and Moynihan 1976, Eriksen 1993, and Banks 1996.

6. These definitions are consistent with those described first in Barth 1969 and refined in Cohen 1978, Banks 1996, and Eriksen 1993.

Chapter 2

1. Historical approaches to the Revolt of 1680 are found in Hackett 1911; Bolton and Marshall 1919; Scholes 1942a, 1942b; Hackett and Shelby 1942; Hammond 1952; Silverberg 1970; Folsom 1973; Espinosa 1988; Knaut 1995; Riley 1995; and Sando 1998.

2. See especially Hodder's discussion of the production of archaeological discourses in *Reading the Past: Current Approaches to Interpretation in Archaeology* (1986: 188). Hodder's discussion of the dialectical relationship between archaeologists and the past references a similar trend in social anthropology. See also Marcus and Fischer 1986, Fabian 1983, and Lowenthal 1985. For a more specific treatment of the other as a creation of archaeological discourse, see Preucel and Hodder 1996: 601–663.

3. See the introduction (ix–xii), as well as Weber's essay, "Reflections on Coronado and the Myth of Quivira," in Weber 1988.

4. See Vizenor's essay "Bone Courts: The Rights and Narrative Representation of Tribal Bones" in Preucel 1996: 652–663, as well as Joan Gero and Delores Root's "Public Representations and Private Concerns: Archaeology in the Pages of *National Geographic*" (Preucel 1996: 531–549).

5. Udall's revisionist stance is complicated by the fact that he served as secretary of the interior during Kennedy's and Johnson's administrations (1961–1969). The Department of Interior oversees the activities of the Bureau of Indian Affairs and has a trust responsibility emanating from treaties and other agreements with Native groups. In recognition of his work in *To the Inland Empire*, Udall was granted knighthood by King Juan Carlos of Spain by the "Order of Isabel the Catholic." See Kessell's foreword in Bolton 1990: xiii.

6. See Las Casas and Zúñiga y Ontiveros 1552.

7. Bandelier's study of Aztec society was instrumental in shaping the theories of social evolution developed in Morgan's *Ancient Society: Researches in the Lines of Human Progress from Savagery through Barbarism to Civilization* (1877). The location of European society as superior to that of Native peoples supported colonial expansion and the replacement of "lesser races" on scientific grounds. See Bandelier et al. 1978 for an analysis of the relationship between the two scholars. Riley and Lange's biographical study of Bandelier (Riley 1996) is an excellent source of information on Bandelier's career, as well as his association with Morgan.

8. Morgan states, "It is both a natural and proper desire to learn . . . how barbarians, by similar progressive advancement, finally attained to civilization;

and why other tribes and nations have been left behind in the race of *progress*—some in civilization, some in barbarism, and others in Savagery" (1877: vi).

9. In 1894, Bolton took courses at the University of Wisconsin Madison from Frederick Jackson Turner. Later, Bolton cited Turner as one of his most important intellectual influences (see Magnaghi 1998: 27–29, 53, 59, 62).

10. Both Bolton and Lummis were instrumental in generating a similar revisionist movement in Spain and in the process supported Franco's own official nationalist narratives. Keen states, "Spanish revisionist historians repeatedly acknowledged the services rendered to their cause by their North American colleagues. Rafael Altamira paid tribute to the contributions of Bourne, Lummis, and Bolton, noting that Spain owed to these historians much of the vindication that had been achieved. Two years after Franco's victory in the Spanish civil war had enthroned the White Legend as the official version of Spain's colonial record, P. Alvarez Rubiano recalled the pioneering efforts of the "North American historical school in tearing down the weak foundations of the Black Legend" (1941: 133).

11. Bannon attempted a similar resurrection of the Catholic image. He characterized the Indian resistance as an "impediment," an inevitable process of Christian conversion and colonial progress. "For long years the chief problems on the Borderlands frontier was the Indian. The Spaniard, after his experience with the Chichimeca on the central plateau, was conditioned to face this *irritant*. He sought to neutralize this vexation with what might be facetiously styled large and frequent applications of the mission formula. And he was reasonably successful despite periodic flare-ups. However, the further north he pushed, the more virulent the irritation became" (124).

12. See Henige 1986: 700–721, 1978: 217–237; Zambardino 1980: 1–27; and Farris 1978: 187–216.

13. Goldstein suggests that the search for "cultural affiliation" and the successful repatriation of Native American materials is complicated by the fact that "successful" research in this area "could well result in the loss of collections to tribes identified as descendents. In other words, *our* success in linking tribes to archaeological cultures might be measured by how much material is repatriated" (1998: 621).

14. For a more thorough discussion of this debate, see the dialogue between J. R. Sackett (1985a, 1985b, 1986, 1990) and Wiessner (1983, 1984, 1985). For discussions of the relationship between style and cultural affiliations, see Conkey and Hastorf 1990; Lemonnier 1989, 1990, 1993a, 1993b; Wobst 1977; and in the prehistoric Southwest, Plog 1978, 1980, 1983, 1990, 1995.

Chapter 3

1. This is not surprising given Radcliffe-Brown's tenure at the University of Chicago during the 1930s. His return to Oxford, Malinowski's tenure at Yale, and Talcott Parsons's attendance of Malinowski's seminars at the London

School of Economics in the 1930s help explain the exchange of ideas between the countries. Parsons also lectured with Radcliffe-Brown at Cambridge in the 1950s, suggesting strong professional and social ties between the two communities of scholars for several decades. See Kuper 1996 for a detailed history of these interactions.

2. This thesis has been advanced elsewhere. See Eriksen 1993: 18–56 and Banks 1996: 24–39 for a discussion of the connections between the RLI and the Chicago school sociologists. Lal's discussion of symbolic interactionsism in Rex and Mason's *Theories of Race and Ethnic Relations* (1986: 281–298) examines the contributions of the Chicago school to ethnic theories using the urban geography of Robert Park.

3. See Park's *Introduction to the Science of Sociology* (1924). In this text, Park advances the tenets of Chicago school sociology; he also anticipates the integration of Simmel's interactionist models with the Freudian emphasis on the individual. The emphasis of individual and personality likewise found expression within the "culture and personality" studies of American anthropology.

4. See Giddens 1984 and Bourdieu 1977.

5. See Mitchell's *Scale of Social Distance* (1958: 23). On the basis of their own perceptions of social distance, subjects were asked what forms of social interaction (choices in marriage, friendship, neighborhood, working partners) they considered acceptable for different groups. Based on social pressures exerted by other members of the group, these categories determined the degree to which groups would interact.

6. Cohen defines these peoples as smaller minorities within the Fulani-Hausa-dominated society. "Hausaness" comprises a single linguistic group, with widely scattered and politically autonomuous communities. Cohen states, "Indeed it is thus possible for some non-Hausa to acquire such roles and thereby become, in effect, Sabo-Hausa. . . . In Sabo a man can qualify as Hausa if he satisfies the following conditions: (1) Speaks Hausa as a first language, (2) can name a place of origin in one of the seven original Hausa states, (3) is a Moslem, (4) has no tribal mark on his face which indicates affiliation to another tribe" (1969: 49). As Hausa identification was precluded by facial scarification (in effect limiting social mobility), facial scarification rapidly declined in these smaller minority communities.

7. Galton's problem refers to auto-correlational sampling errors in statistics; to ensure statistical validity, traits selected for comparison must originate in separate domains.

8. In his monograph on the Kachin in Upper Burma (1954), Leach demonstrated a high degree of fluidity within a single, multi-"cultural" system. There are multiple, sometimes antagonistic social spheres in which individuals participate. This ambiguity provides the context in which power and status are proffered via the group on the individual. Leach does (obliquely) acknowledge the influence of Parsons (10) and Merton (fn. 13) for this idea.

9. More recently, Barth has suggested a similar position and attributes much of the ink spilled in these debates as largely fruitless. He suggests micro, macro,

and median as useful levels of analysis. See his essay in Vermeulen and Govers 1994: 11–31.

10. For specific case studies, see Brass's (1985) discussion of elite competition within ethnic movements, Horowitz's (1985) *Ethnic Groups in Conflict.*

Chapter 4

1. In *Nations and Nationalism*, Gellner (1983) characterizes nationalist identities as the manifestation of *false consciousness*: "Generally speaking, nationalist ideologies suffer from a pervasive false consciousness. Its myths invert reality . . . it claims to protect an old folk society while in fact helping to build a up an anonymous mass society" (124).

2. The legacy of these social hierarchies continues to inform movements for social reform in Mexico today. See Fuentes 1992.

3. The literature regarding this debate is extensive. Hanke's contributions (1941, 1970, 1974, 1979) to the philosophical debates regarding the place of the Indians in Spanish society are still considered to be the definitive works on the subject. See also Sullivan (1995) for a review of the literature. Gellner's characterization of ethnic nationalism is helpful: "Nationalisms are . . . ethnic ideologies which hold that their group should dominate the state. A nation-state, therefore is a state dominated by an ethnic group whose markers of identity (such as language or religion) are frequently embedded in its official symbolism and legislation" (1983: 99).

4. Eriksen states, "Vernacularisation is an important aspect of many nationalist movements, since a shared language can be a powerful symbol of cultural unity as well as a convenient tool in the administration of a nation state" (1993: 103). Antonio de Nebrija, the architect of Spain's colonial empire, understood the importance of language in empire building. In a discussion with Queen Isabella, Nebrija was reported to have stated that "language is the perfect instrument of empire" (Elliott 1969: 128).

5. For an extended discussion of this thesis, see Pagden 1993.

6. The full statement is as follows:

> One of these Pontiffs, who succeeded St. Peter as Lord of the world, in the dignity and seat which I have before mentioned, made donation of these said Isles and Terra-firma to the aforesaid King and Queen and to their successors, our lords, with all that there are in these territories, as is contained in certain writings which passed upon the subject as aforesaid, which you can see if you wish.
>
> So their Highnesses are kings and lords of these Islands and land of Terra-firma by virtue of this donation; and some islands, and indeed almost all those to whom this has been notified have received and served their Highnesses, as lords and kings, in the way that subjects ought to do, with good will, without any resistance, immediately, without delay, when they were informed of the aforesaid facts. And also they received and obeyed the priests whom their highness sent to preach to them and to teach them our Holy Faith; and all these of their own free will, without any reward or condition, have become Christians and are so, and

their Highnesses have joyfully and benignantly received them and also have commanded them to be treated as their subjects and vassals; and you too are held and obliged to do the same. Therefore, as best we can, we ask and require that you consider what we have said to you, and that you take the time that shall be necessary to understand and deliberate upon it, and that you acknowledge the Church as the Ruler and Superior of the whole world and the high priest called Pope, and in his name the King and Queen Doña Juana, our lords, in his place, as superiors and lords and kings of these islands and this Terra-firma, by virtue of the said donation, and that you consent and give place that these religious fathers should declare and preach to you the aforesaid.

If you do so, you will do well, and that which you are obliged to do their Highnesses, and we in their name shall receive you in all love and charity, and shall leave you your wives, and your children, and your lands, free without servitude, that you may do with them and with yourselves freely, that which you like and think best, and they shall not compel you to turn Christian, unless you yourselves, when informed of the truth should wish to be converted to our Holy Catholic Faith, as almost all of the inhabitants of the rest of the islands have done. And besides this, their Highnesses award you many privileges and exceptions and will grant you many benefits.

But if you do not do this, and wickedly and intentionally delay to do so, I certify to you that, with the help of God, we shall forcibly enter into your country and shall make war against you in all ways and manners that we can, and shall subject you to the yoke and obedience of the Church and their Highnesses: we shall take you, and your wives and your children, and shall make slaves of them, and as such shall sell and dispose of them as their Highnesses may command; and we shall take away your goods, and shall do all the harm and damage that we can, as to vassals who do not obey, and refuse to receive their lord, and resist and contradict him; and we protest that the deaths and losses which shall accrue from this are your fault, and not that of their Highnesses, or ours, nor of these cavaliers who come with us. And that we have said this to you and made this Requisition, we request the notary here present to give us his testimony in writing, and we ask the rest who are present that they should be witness of this Requisition.

7. For a comprehensive discussion of Spain's financial policies and economic crises, see 1963: 199–200, 210–231, 263–290, 333–367.

8. An extensive literature deals specifically with the legal conditions under which the enslavement of Indians was permitted. For specific studies of Spanish legal policies, see Williams 1990, Elliot 1969: 56–74, Hanke 1974, and Lewis 1995. Patterson's *Slavery and Social Death: A Comparative Study* (1982) and Davis's *The Problem of Slavery in Western Culture* (1966) are also classic texts dealing with the history of slavery in the Americas.

Chapter 5

1. Both Knaut (1995) and Scholes (1942a, 1942b) make this argument.

2. Riley (1971: 285–314) suggests that Nahuatl and Tarascan speakers were available for Coronado's initial journey and that Lower Pima speakers likely

facilitated communication between the colonists and Pueblos. After the pacification of the Zunis, Coronado also mentions the presence of people from Cicuique at Zuni ("*Bigotes*" from Pecos) who had been informed of the presence of the Spanish and come to "make their acquaintance and become friends" (Hammond1940: 324–325). In subsequent reports of Espejo's entrada, Espejo claims to have met three Indians who were left from Coronados's expedition and who spoke Nahuatl (Hammond and Rey 1966: 225). Obregón's *Historia* (1584) similarly documents the presence of two brothers named Andrés and Gaspar who remained at Zuni after Coronado's departure (Hammond and Rey 1928: 292). In addition, Espejo's chronicler Diego Pérez de Luxán states that Zuni were descendants of those left by Coronado's Nahuatl and Tarascan escorts (Hammond 1966: 184). In 1596, nearly sixty years later, Oñate similarly mentions "descendents of the Mexican Indians left at Zuni . . . by Coronado."

3. "In the year 1530, when Nuño de Guzmán was president of New Spain, he had under his authority an Indian whom the Spaniards called Tejo, from the valley or valleys of Oxitipar. This Indian said that he was the son of a trader who was dead and declared that when he was very small, his father used to go into the interior of the land to trade rich colored plumes, used for feather crests, and in exchange brought back large quantities of gold and silver." The journey was said to have taken "forty days of travel over entirely deserted country . . . to the north, through the country between the two seas" and that the cities were "so large that he would compare them with Mexico City or its surroundings" (Hammond 1940: 95).

4. "The requisition . . . made intelligible to the people of the country through an interpreter. But they, being a proud people, paid little attention to it" (Hammond and Rey 1940: 168).

5. See Barrett 1997: 234.

6. At Captian López de Cardenas's trial, he stated that Coronado, charged with the forcible removal of the Tiwas (among other crimes), denied these charges, stating, "The Indians gave everything willingly" (Hammond and Rey 1940: 330).

7. Castañeda narrates, "As soon as a Spaniard came to the Pueblo, he demanded the supplies at once, and they had to give them. . . . With all this there was nothing the natives could do except take off their own cloaks until the number that the Spanish had asked for was reached" (Hammond and Rey 1940: 224).

8. Coronado denied that the incident had occurred or that it had led to the conflict, although Castañeda explains the incident at some length (Hammond and Rey 1940: 224–225, 330). In an earlier testimony, Coronado's horseman, Joan de Contreras, identified the man as Juan de Villegas, the brother of a *regidor* of Mexico City. As a result of this association, Villegas was never tried for the crime. See "Coronado's Testimony" in Hammond (1940: 330, fn. 17).

9. In the official inquiry, the Juez de Residencia of New Galicia, Lorenzo de Tejada heard reports that "he and the Spaniards who went with him committed, in going and returning, great cruelties against the natives of the lands

through which they passed, killing large numbers of them and committing other acts and injustices to the detriment of the service of God and ours" (313).

10. The sieges at Moho took place between December and March. According to Castañeda's *relacion*, the rivers and ground near the pueblos was still frozen at this time of year, raising questions about Coronado's account of the Pecos' response (Hammond and Rey 1940: 230).

11. There has been considerable attention devoted to the identity of the Teyas. Riley (1997: 320–43) provides an extensive discussion of the debate as well as the current consensus regarding their identity.

12. The identities of the Tutahaco pueblos have never been clearly determined (see Hammond and Rey 1940: 220, fn. 1).

13. Coronado was apparently concerned for his own safety. After meetings were held in his absence, Coronado's men urged him to return, "since they had not found any wealth, nor had any settled area been discovered where *repartamientos* could be provided for the whole army" (Hammond and Rey 1940: 265, 267). Before leaving, Coronado's men "collected" servants from among the Pueblos but were forced to release them before they returned south. Other, unnamed Indians from his party remained behind it Cibola (271).

14. Bolton states that Lorenzo de Tejada, a judge of the Royal Audiencia of Mexico City, received his commission for the investigation of Coronado on "the same ship that brought Tello de Sandoval, Visitador General" who came to investigate Mendoza and enforce the New Laws regarding Indian Labor and tributes" (1949: 368). During the trial, witnesses from the expedition were called and cross-examined behind closed doors. Coronado's testimony (Hammond and Rey 1940: 319–336, 369–397) reveals several discrepancies with the subsequent testimony of Cárdenas (337–368) as well as the account related by Casañeda (191–283). When asked about the burnings of the people who surrendered at Arenal and Moho, he stated that "he did not know that such a thing had been done" (334). Coronado claimed that he was completely unaware of any grievances the Pueblos had and "without having given them any cause that this witness knows of, they revolted" (331). When the commission received reports that Coronado had ordered Cárdenas "not to burn them until the arrival of the Indians named Turk and Isopete in order that they might witness it and spread the report in their countries of the justice that was meted out to those who rebelled," he stated that he "had never heard" of the incident, "nor had any such information ever reached him that he remembers. . . . He heard that the Indians in the tent had revolted or offered resistance and that some of them had been killed" (335). He was found guilty of attacking Hopi, Zuni, and Pecos and causing the Tiguex rebellion, but these charges were later reversed in Mexico City by the Audiencia of Mexico City (Hammond and Rey 1940: 393–398).

15. Even use of the term *conquest* was outlawed. Discoveries were to be undertaken "with peace and mercy" (Hammond 1966: 6).

16. The full citation is as follows:

> We gave them to understand through an interpreter that we, who were now in their land, had come only to restore friendship between them and their enemies,

to help them in their wars and struggles, and to lend them the protection and aid they might need against their foes. We also told them not to fear the Spaniards, who would not cause them any further harm; and that we were there for this express purpose—to insure that in the future no Spaniards would come except as their friends, provided the Indians behaved well, but if they misbehaved, the Spaniards would kill them all. If they wanted to avenge the seizure of their friends, relatives, wives and children, they should come out into the open quickly; for we eight, who were there among them, would avenge the other Spaniards for what they had suffered. Our boldness towards the Natives was meant primarily to intimidate them, so that the news should spread. We fired Quite a few harquebus shots, at which the natives were very much frightened. (Hammond 1966: 71)

17. Both Hammond and Schroeder identify this as the Keres (or possibly Tewa) settlement of San Marcos (1966: 13; Schroeder 1979: 244–247). See Dozier for the identification of San Marcos as a Tewa settlement (1970b: 39–43).

18. The identity of Piedra Aita is not definitively known. Schroeder (1979: 248) speculates that this was San Cristóbal, but Reed argues that Piedrahita was Pecos (Reed 1943: 254–264, 276–288).

19. Unfortunately, none of the clergy or Chamuscado, the party's leader, survived to provide an alternate account of any of these events. It is possible that tensions between Chamuscado, Gallegos, and the Franciscans over the activities and objectives of the party led to the departure of the missionaries; given the inconsistencies of Gallegos's account of the departure, it is possible that other events (not recorded in the official transcripts sent to the viceroy) led to their demise.

20. Puaray has been identified by Bandelier as being two miles south of Kuaua, but the exact location of the pueblo is unknown (see Bandelier 1892: 233).

21. As far as Gallegos's list of Pueblos, there are significant differences between this list and a separate document referred to as "Pedrosa's List of Pueblos" (Hammond 1966: 115).

22. Antonio Espejo had been jailed and released in 1578 for stealing cattle. Wanted for these charges, Antonio had murdered a ranch hand in 1580, and he and his brother had killed an Indian ranch hand in a separate incident in 1581. Both were brought to trial in Mexico City for the second murder, but through their association with Inquisition officials, they were not severely punished (Conway 1931: 1–20). Antonio was fined and escaped to the frontier settlement of Santa Barbara where he learned of the returning expedition (Hammond 1966: 16–17).

23. Hammond suggests that this pueblo, named Maguas or Magrias, was a Tompiro pueblo (1966: 174, fn. 45).

24. See Hammond 1966: 176, fn. 46.

25. See Hammond 1966 (142, 184, 226) for Luxán's and Espejo's respective figures.

26. Narrated by Luxán, neither Espejo nor Obregón mention this incident in their accounts (Hammond 1966: 204, fn. 118)

27. According to Hammond, "Castaño de Sosa . . . sent out Luis Bogado and other Captains on new raiding parties to capture Indians for the slave markets of northern Mexico this practice was in conformity with the old policy of the frontier, where aggressive captains supplied the mines and ranchos with labor in this manner. . . . This activity not only served to keep the men occupied, but to provide them with servants as well as with compensation for their labor and investment in making such an expedition, and thus to prevent discouragement and desertion (1966: 31). Governor Carvajal was later arrested by the viceroy for "establishing" bogus settlements and collecting subsidies from these for the Crown. Villamanrique narrates:

> Now I learned that while he was in prison he empowered a certain Gaspar Castaño de Sosa to act as his lieutenant, and named various other aides, all of whom are following in his footsteps. They are stationed at the above mentioned locality of Almaden in Cuala, with more that sixty soldiers—outlaws, criminals and murderers—who practice neither justice nor piety and are raising a rebellion in defiance of both God and king. These men invade the interior, seize peaceable Indians, and sell them in Mazapil, Saltillo, Sombrerete, and indeed everywhere in that region (297).

28. Even the king was notified of the nature of the slave trade in the region:

> In a letter to your Majesty . . . I reported in detail on the situation of Governor Luis de Carvajal . . . as well as on the efforts of a certain Gaspar Castaño (whom he had appointed as his lieutenant), in conjunction with the vagabonds who had joined him and indeed all the riffraff left over from the war against the Chichimecas, to penetrate the interior with the purpose of exploring New Mexico. . . . I also notified you of the harm that these persons were doing to the natives by entering into their lands to seize and enslave them, though the victims gave no cause and were blameless, so that the whole kingdom was in turmoil. . . . My purpose has been to punish the disobedient . . . and also to prevent the disturbance and unrest aroused among the natives by their enslavement, so that they may be persuaded that we desire only their well-being and conversion, without seeking anything further from them. (Hammond 1966: 299–303)

In a letter from King Phillip II to the audiencia of New Spain (dated January 17, 1593), the king instructed the audiencia that "you shall neither encourage nor permit the enslavement of any Indians whatsoever, and if you hear of any who have been made slaves, you shall set them free, with proclamations by the town crier throughout that city and in Guadalajara, New Galicia" (317).

29. Convicted by Velasco and exiled to the Philippines, Castaño was killed in a rebellion of Chinese galley slaves aboard a Spanish vessel (Hammond 1966: 48). In his testimony before the viceroy, Castaño defended his actions among the Pueblos and explained the economic necessities leading to the slave trade on the frontier: "if I have indeed erred, I did so in sincere reliance upon authority granted by . . . Luis de Carvajal. . . . It was in the exercise of his authority, acting as his lieutenant, that I did the things which must have been reported to your Grace." Later, Castaño de Sosa recounts his journey to the pueblos where, "after all these trials, God was pleased to lead me to a region where I found

many pueblos. Among these settlements I discovered some mines, which was my chief aim. Moreover, during that short time, by kindliness and affection, I induced the natives to pledge obedience to the king, and they received from me the protection which was their due, in conformity with his Majesty's commands" (Hammond 1966: 307) Later, he defends his abandonment of the mines and farms of New Leon: "All of these things were lost because there was no silver, and the land was such that we could not support ourselves without resorting to acts which would have been in a sense wrongful. . . . The few that I did take were seized in order to punish some culprits who had slain three of my Spaniards and committed a thousand other outrages and injurious acts" (309).

30. In a letter from Viceroy Velasco to King Phillip (dated February 28, 1592), Velasco communicates the possibility of extracting wealth from the Pueblos:

> I felt and still feel that it was most fitting and judicious to remove Castaño de Sosa and his soldiers from that region; yet I also believe that it would be fitting and judicious to undertake the pacification of the natives there. The attempt might prove worthwhile, quite aside from the service that would be rendered to our Lord through the conversion of so many natives, because they are numerous and the land has a good climate, is fertile, and produces abundant food supplies. . . . It is readily apparent that no one will care to enter into a contract for this venture without assurance of great advantages and profit, or without the aim and prospect of encomiendas and tribute from the Indians. Since the principal objective should be the conversion of these natives by proper means, whereby all the other results (which may, indeed be materially advantageous) are to be justified.

31. He requested, "That I and my successors be authorized to levy the tributes which the Indians will have to give, according to the fruits of their land, not only the tribute of the King our lord, but also that for the other encomenderos, with free and absolute authority to raise and lower them or to commute them in any way that may seem fitting to me or my successors, without reference to the Viceroy or Audiencia" (34).

32. Oñate's muster lists (among other things) over 200 horses, 200 oxen, 1,000 goats, 4,300 sheep, 1,400 cattle, clothing, medicines, wheat, corn, armor, and weapons. Many of the livestock escaped during the journey north (Hammond and Rey 1953: 199–308).

33. "Doña Inés . . . is the Indian woman we brought from Mexico like a second Malinche, but she does not know the language or any other spoken in New Mexico" (Hammond 1953: 321).

34. "The Indians, suspicious and excited, had abandoned the Pueblo" (Hammond 1953: 318).

35. The author narrates, "We found people only at this pueblo, and at the first and second; all the others we found deserted" (Hammond and Rey 1953: 318).

36. "Among them, and they seemed like spies, was the one whom we called Don Lope, sent by Tomás and Cristóbal, Indians who had remained there since the time of Castaño" (319).

37. This settlement was partially excavated in 1959 by Florence Hawley Ellis (Ellis and Florence Hawley Ellis Museum of Anthropology 1989).

38. According to Velasco, tributes were collected: "Against the wishes of the natives and to their great grief . . . we had to support ourselves by taking as much as we could from each one. . . . The system employed during this time to feed more than five hundred persons . . . has been to send people out every month in various directions to bring maize from the pueblos. The feelings of the natives against supplying it cannot be exaggerated. . . . They weep and cry out as if they and all of their descendents were being killed" (Hammond and Rey 1953: 610).

39. See "Letter from the Council of the Indies to the King, July 7, 1602" (Hammond and Rey 1953: 972).

40. This figure is difficult to imagine. Barnes, Naylor, and Polzer (1981: 73) have identified approximate equivalencies for Spanish units of measurement. Based on these measures, each pueblo would have had to provide *twenty thousand* pounds of corn per tribute period.

41. This starvation and treatment of the Pueblos were confirmed (in separate testimonies) through letters written by the colonists to the viceroy from San Gabriel. See "Investigation of Conditions in New Mexico" (Hammond 1953: 623–689).

42. According to the "Letter from Captain Velasco to the Viceroy, March 22, 1601," "Equally cruel was the war recently waged against the Jumanos. It was caused by similar attempts to take their food and blankets. The sargento mayor, with seventy men, waged war against them by fire and sword, killed more than nine hundred Indians, burned and leveled their pueblo and took more than two hundred prisoners" (Hammond and Rey 1953: 615; for corroborating testimony, see 650–651, 693).

43. The full text is more instructive: "I cannot help but inform your majesty that this conquest is becoming a fairy tale. If those who write the reports [Oñate] imagine that they are believed by those who read them, they are greatly mistaken. Less substance is being revealed every day. . . . There is no end to this business, because they keep expanding it by passing from one undertaking to another" (1009).

44. Oñate was found guilty of abusing both his soldiers and the missionaries and "that he abused and permitted others to maltreat the natives of the said provinces, wherever he passed with his people, taking away from them by force their corn and whatever else they had, thereby discrediting the motives for their pacification and his own commission" ("Conviction of Oñate and His Captains, 1614," in Hammond and Rey 1953: 1109–1110).

45. Based on the personal letters and official church documents drawn from Catholic and Spanish archives throughout Mexico and Spain, Scholes's work remains the definitive analysis of this poorly known period. More recent studies such as Gutiérrez's *When Jesus Came, the Corn Mothers Went Away* (1991), Kessell's *Kiva, Cross and Crown* (1979), and Knaut's *Pueblo Revolt of 1680* (1995) are likewise excellent secondary sources.

46. For a more detailed discussion of these negotiations, see Ivey 1998: 121–152.

47. For an account of a similar event at Pecos, see Scholes 1936: 169, fn. 15.

48. Eulate's tributes were more extensive than any of his predecessors'. The governor was able to send nine wagon loads of pinyon, hides, and mantas south to the mining districts of Parral (Scholes 1936: 287).

49. Ten of the replacements sent north with Mendízabal had deserted, and ten others had died serving in the colony (Scholes 1930: 199).

50. López heard complaints against Antonio de Salas, encomendero of Pojoaque, by several Tewa villages. In what seems to have been a fairly common occurrence, Salas's herds had grazed on Pueblo crops and violated prohibitions against the building of colonial settlements within the pueblos proper. López demolished Salas's home, citing the regulations written by the Crown (Scholes 1942a: 43, 63–78).

51. For a list of the reputed activities of the friars, see Scholes (1942a: 89, 65–73, 83, 99). For discussions of accusations made by Franciscan missionaries that López (as well as his wife) were "crypto-Jews" (a term applied to "new Christians" in Spain), see Scholes (1942a: 83, 99).

52. López died in prison in 1664, while defending himself before the Inquisition. The testimony of him and his wife demonstrate the level of suspicion the governor and his wife were regarded with (Scholes 1942a: 149–170). Both were cleared of the charges. Several of López's allies were similarly jailed and forced to defend themselves before Inquisition officials. Some were exonerated, while others were sentenced to serve the government in various capacities for limited tenures. Generally, their fates are not known, although some may have returned to their families in New Mexico (Scholes 1942a: 172–197). Given the animosity they likely held toward the friars, it is interesting to speculate as to whether they returned to New Mexico or played any role in the Revolt of 1680. To date, no testimonies supporting this notion have come to light.

53. "For three years now no crops have been harvested. In the past year, 1668, a great many Indians perished of hunger, lying dead along the roads, in the ravines and in their hut. There were pueblos where more than 450 died of hunger." From "Letter from Fray Juan Bernal, to the Tribunal, April 1669, from the Convento de Santo Domingo," in Hackett 1937: 272–273.

54 See Jones's (1966) discussion of Puebloan auxiliaries in New Mexico.

55. See Gutiérrez 1991 (197, 289–292) for a discussion of these intermediate social categories, exogamy, and illegitimate birthrates.

Chapter 6

1. After the rebellion, Franciscan missionaries noted, "The Indians who have done the greatest harm are those who have been most favored by the religious and who are most intelligent" (Hackett and Shelby 1942: 59).

2. Spanish and English names for canyons, valleys, and rivers can be difficult to correlate with historical references; English and Spanish names are often used interchangeably in archaeological literature. Place names referred to in this chapter are therefore consistent with those used on U.S. Geological Survey (USGS) quadrants. The Museum of New Mexico's Laboratory of Anthropology, Forest Service archaeologists, and tribal cultural resource officers predominantly use USGS quads to catalog and locate sites for regional databases.

3. See Wills 1990 (319–322) for an examination of various theoretical approaches to the introduction and development of agriculture in the Southwest.

4. A similar situation exists on many U.S. Forest Service, Bureau of Land Management, state, and private lands. Where surveys and excavations have been completed, these reports, often of highly variable quality, are often difficult to access and locate. In many cases, the most accurate information regarding archaeological sites and recent research is informal interviews with archaeologists, tribal historians, and preservation officers who work in cultural resource management offices. Often the lag time between the preliminary findings of field projects and the publication of information in academic literature poses a significant obstacle for any kind of regional settlement analysis.

5. Following the creation of the Coronado Cuarto Centennial Commission by the New Mexico State Legislature in 1935, Senator Clinton P. Anderson secured a $5 million congressional appropriation for the creation of the United States Coronado Exposition Commission (Zimmerman 1940: 158–162). According to Kessell (1989: xxi), the series was published to strengthen the historical connections between the United States and Mexico during World War II (see also Bannon 1978: 220–221). The Coronado National Monument, now located south of Bernalillo, was originally to have straddled the U.S.-Mexican border in Arizona in a symbolic act of historical convergence. The multivolumed Coronado series was to be published over several years by several of Bolton's students; but with the exception of Reiter's abbreviated historical analysis provided by Frances Scholes, there appears to have been little sustained collaboration between historians and archaeologists during this period. The timing of the various publications may have been part of the problem. While the entire series covers New Mexican history from the sixteenth to the eighteenth centuries, the volumes were published out of sequence: Bolton's *Volume I: Coronado on the Turquoise Trail, Knight of Pueblos and Plains* was not completed until 1949, and Hammond and Rey's *Volume III: Rediscovery of New Mexico, 1580–1594* was not published until 1966. Other volumes, such as Frances Scholes's coverage of seventeenth-century history (Vol. VII), were never completed.

6. This view is still not widely accepted by archaeologists. For example, in the more geographically isolated Pueblo settlements such as Zuni, there is little evidence supporting a militarily driven need for aggregation (Kintigh 1996: 131–144); in the Upper Rio Grande, there is more evidence supporting

reciprocal relationships between the Pueblos and their Athapaskan neighbors (Spielmann 1983: 257–272, 1991a, 1991b).

7. The School of American Research sponsored sporadic excavations and material collection at Jemez sites in 1909, 1910, 1911, 1912, 1914, 1921, and 1922. Euphemistically referred to as "reconnaissance" by Reiter (1938: 7), none of the material was (or has ever been) systematically studied, nor has any of this work been published.

8. For more detailed discussions of San Gabriel, see Bandelier 1890: 311 and 1892: 31; 36, fn. 1; 58; 59; 123; 333, fn. 1.

9. Kivas and other ceremonial structures (medicine society houses, etc.) can take any number of shapes (Adams 1991).

10. Information dealing with the specific locations of these sites as well as more detailed maps have been purposely deleted from this section. Interested parties should contact the author or Forest Service personnel for access to these materials.

11. A Girl Scout troop was allowed to perform excavations at the site in 1968 under the sponsorship of the Museum of New Mexico. The notes from this excavation, on file at the Museum of New Mexico's Lab of Anthropology, emphasize the clear historical affiliation between the Jemez and Boletsakwa. They also demonstrate the complete lack of communication between archaeologists and the known descendants of Boletsakwa. A trench was excavated on the downslope (eastern) side of roomblock A where a human burial had eroded from the site. "Permission was received to sink an exploratory trench through the wall debris up to the wall. . . . Human and animal bone fragments . . . were recovered from the trash wall debris." As a result of this often cavalier approach to excavation among archaeologists and museum personnel, excavations on Forest Service lands are highly restricted. (See Museum of New Mexico files, LA 136.)

12. Elliot's work is a reprint of National Park Service Registry information on file at the Lab of Anthropology in Santa Fe. Boletsakwa was recorded in a sketch map drawn by H. P. Mera in the 1930s. W. S. Stallings acquired tree ring samples from the site at the same time, although there are no extant field notes from either of these field projects. Excavations must have taken place during Stallings's visit, as there are no visible or exposed sources of roof beams at the site. Since then, the main village has been heavily disturbed by pot hunters.

13. Hausley's thesis, "Opuntia Imbricata Distribution on Old Jemez Indian Habitation Sites," is a primarily a botanical study of cholla cacti on archaeological sites.

14. Bandelier correctly identified the site as Kwastiukwa in his *Final Report, Volume II* (1892: 205–206). Later, Wesley Bradfield conducted excavations at Amoxiumqua (LA 481) but incorrectly identified the site as Kwastiukwa. Adding to the confusion, they began work in 1914 at "the site west of Virgin canyon" (Reiter 1938: 81), now known to be Kwastiukwa (LA 482), and misidentified the site as Amoxiumqua. Working with the Bureau of American Ethnology, the School of American Research, and the Royal Ontario Museum,

Bradfield, working under the direction of Charles F. Lummis, conducted test excavations in 1914 at Kwastiukwa. The results of this excavation are still visible in the northeastern portion of roomblock F, where seventeen to twenty rooms were completely excavated. The notes from both of these excavations have never been located. The status of the artifact collections gathered from this effort are likewise unknown. Adding to the confusion, both Harrington (1916: 395) and Hodge (1916: 242, fn. 32) similarly misidentify Amoxiumqua as Kwastiukwa. Since then, Kwastiukwa has seen a resurgence of activity. In 1988, Winifred Creamer and Jonathan Haas conducted limited excavations in plaza 3, just south and east of kiva 1, as well as to the east of roomblock F. In 1989, William Whatley of Archaeological Research-Exploration initiated a mapping and detailed survey project in collaboration with the Pueblo of Jemez and U.S. Forest Service (Jemez Ranger District) archaeologists Tom Cartedge and Carol Raish, the Kwastiukwa Interdisciplinary Archaeological Research Project. The results of this project have been used in educational programs at the pueblo, but they have not yet been made public in Forest Service or Museum of New Mexico, Laboratory of Anthropology, files.

15. An artist's rendition of the site, available at the Walatowa (Jemez) Cultural Center (apparently produced from William Whatley's survey in 1988), depicts individual roomblocks containing four to five stories.

16. These rooms were most likely those excavated by Bradfield in 1914.

17. F. W. Hodge and Harrington (1916) confused Amoxiumqua (LA 481, FS 530) with Kwastiyukwa (LA 482 or FS 11), also known as "Giant Footprint Ruin." Their initial field seasons (1910, 1911, 1914) were spent at LA 481 (the actual Amoxiumqua), but their field notes mistakenly identify the site as "Kwastiukwa." In 1914, work was begun at "a site to the west" (Reiter 1938: 81) which Hodge and Harrington misidentified as Amoxiumqua. Reiter provides the most detailed summary of the work completed at Amoxiumqua as well as the confusion surrounding the identification of the site by Harrington.

18. Amoxiumqua was partially excavated by several organizations. Collections of materials are located at the Maxwell Museum, the Laboratory of Anthropology in New Mexico, as well as the Royal Ontario Museum of Archaeology in Toronto. The Bureau of American Ethnology (BAE) also sponsored expeditions to collect materials from the site. These are presumably housed at the Smithsonian Institute. According to Reiter, Kenneth Chapman (1911) made collections on behalf of the BAE at several of the large sites in the Jemez region. F. W. Hodge, Jesse Nusbaum, and Chapman excavated rooms and burial mounds at Amoxiumqua over four seasons. Unfortunately, no publications resulted from this work. The only published material relating to these excavations is Reiter's discussion of "earlier fieldwork" in chapter 4 (81–92). Hodge and Chapman worked at the site in 1910 and 1911 under the auspices of the BAE. In 1912 and 1914, Wesley Bradfield and Charles F. Lummis continued the excavations, working this time for the Royal Ontario Museum. According to Reiter (81), Bradfield's field notes from the excavations were "mislaid," and no mention is made of where the collections produced from the excavations were located.

19. Holmes's (1905) work on Amoxiumqua was literally copied word for word in Hewett's 1906 publication, although neither seems to have been overly concerned with citing the sources of their data. Where figures concerning measurements and estimates of kivas, site size, and other feature size estimates are present, one must assume that Holmes's earlier publication should be credited with the fieldwork. It may have been that either Bandelier's notes or his final report of 1890 were consulted by both individuals.

20. H. P. Mera and Reginald Fisher made maps of the site on file at the Museum of New Mexico's Lab of Anthropology. Mera mislabeled "Tovakwa" (LA 483) in his sketches as LA 484.

21. Mera's sketch map, drawn in the 1930s, clearly shows kiva 5 in the northwest corner of the plaza. Ramenofsky's mapping project located the kiva in the center of section 3C. Erosion and "wall fall" may account for this discrepancy.

22. Bandelier seems to have been confused about which of the sites he visited were of prehistoric origin. Apparently, he was under the impression that Tamaya (LA 2049) was a site that predated the contact period; this is understandable given the fact that "Tamaya" refers to a general location for Santa Ana peoples and not necessarily of a specific community (Lange 1975: 69). Bandelier (1892: 196) mentions that informants had claimed that the site was among three named Tamaya (the contemporary name for Santa Ana) that are ancestral to the Keres-speaking people of Santa Ana. Based on historical documents, Bandelier suggests that the site was erected previous to Oñate's 1598 entrada and was destroyed by Posada's failed attempt to reconquer the region in 1687 (1892: 193). He also mentions that the governor of Santa Ana stated that the people of Santa Ana had moved from a village to the south of the modern pueblo to the mesa. Later they returned to the pueblo at the present location (Lange 1975: 69). Regardless of which site was attacked, the people of Santa Ana were targeted by Pedro Reneros de Posada, governor of El Paso del Norte (1686–1688). Refusing to surrender to Posada, the town was assaulted, and "the town was captured an burned. A number of Indians were killed in the fighting and in the burning of the pueblo which followed the Spanish victory" (1892: 194–195). Twitchell cites a proclamation relating to the attack dated October 6, 1687, "sentencing ten Queres captives to be sold into slavery in Nueva Vizcaya for ten years" for resisting the Spanish (1914: 76).

23. Warren's analysis did not include petrographic analysis, and a more detailed analysis of the samples may reveal alternative explanations of source materials, particularly the "Zia Basalt" referenced in Warren's analysis.

Chapter 7

1. Utctsityi in White 1940: 85; Uretsiti in Benedict 1931: 1.

2. "Naotsity felt superior to her sister and challenged her to a number of contests so that she could demonstrate her superiority" (White 1940: 85).

3. In one particularly telling contest, Uretsiti defeats Naotsiti by catching the first rays of sun before her taller sister. "'Why is it that you are standing covered

with sun and I am still in shadow?' Uretsiti answers, 'It is not I who have done this, it is our Mother. This was your challenge, but she has given me power to overcome. . . . Just as the Mother of the Pueblos was safe and won in this contest, so whenever there is fighting between the Navajos and the Pueblos, the people of the Pueblos are safe and win'" (Benedict 1931: 1–4; see also Parsons 1939: 244–247).

4. "Many people came to this country and wanted to settle there. Nan'kortez came with many soldiers. He went among the Pueblos and did great damage. He tried to end all customs and religions" (Benedict 1931: 191–192).

5. Bandelier, born in Switzerland, was apparently fluent in English, French, German, and Spanish, but the accuracy of his Spanish translations has been questioned by some scholars (Lange and Riley 1966: 8-8, fn. 2).

6. Letter from Clark Wissler to Nels Nelson, dated May 9, 1912, American Museum of Natural History (AMNH) archives.

7. Letter from Nels Nelson to Clark Wissler, dated November 19, 1912, AMNH archives.

8. It is also unclear whether his informants were from the pueblo itself. Workers at the site were drawn from "Mexicans" he met through a Spanish resident of the pueblo. The excavation of the site was most likely noticed by members of the pueblo, and his failure to contract workers from the pueblo indicates a general reluctance on the part of the Pueblos to endorse the project (Preucel et al. 1996: 24).

Chapter 8

1. For a more extensive discussion of genízaro populations, see Adams and Chavez 1956: 42, 252–256; Jones 1966: 36, 138–142; Dozier 1970b: 84–86; Gutiérrez 1991: 148–156, 171–196, 295–297.

References

Adams, E. Charles

1991 *The Origin and Development of the Pueblo Katsina Cult.* Tucson: University of Arizona Press.

Adams, Eleanor

1954 Bishop Tamaron's Visitation of New Mexico. *Historical Society of New Mexico Publications in History* 15:1–109. Albuquerque: University of New Mexico.

Adelman, J., and S. Aron

1999 From Borderlands to Borders: Empires, Nation-States, and the Peoples in Between in North American History. *American Historical Review* (June):814–841.

Adler, M. A.

1994 Population Aggregation and the Anasazi Social Landscape: A View from the Four Corners. In *The Ancient Southwestern Community: Models and Methods for the Study of Prehistoric Social Organization*, edited by W. H. W. and R. D. Leonard. Albuquerque: University of New Mexico Press.

Adler, M. A., and A. J. Adler

1996 Mapping the Puebloan Southwest. In *The Prehistoric Pueblo World: A.D. 1150–1350*, edited by M. A. Adler. Tucson: University of Arizona Press.

Adler, M. A., Todd Van Pool, and Robert D. Leonard

1996 Ancestral Pueblo Population Aggregation and Abandonment in the North American Southwest. *Journal of World Prehistory* 10(3):375–438.

Alexander, H. G., and P. R. Alexander

1935 *Report on the Excavation of Jemez Cave, New Mexico.* Monograph of the School of American Research. Santa Fe, NM: School of American Research Press.

Anderson, Benedict

1983 *Imagined Communities: Reflections on the Origin and Spread of Nationalism*. London: Verso Press.

Asad, T.

1973 *Anthropology and the Colonial Encounter*. London: Ithaca Press.

Ashcroft, B., G. Griffiths, and H. Tiffin

1995 *The Post-colonial Studies Reader*. London: Routledge Press.

Atalay, Sonya

2006 Indigenous Archaeology as Decolonizing Practice. *American Indian Quarterly: Special Issue: Decolonizing Archaeology* 30(3): 269–279.

Ayer, E. E.

1916 *Memorial of Fray Alonzo de Benavides, 1630*. Chicago: Private printing.

Ayers, J. E.

1995 The Archaeology of Spanish and Mexican Colonization in New Mexico. *Historical Archeology* 3:53–68.

Bailey, J. B.

1940 *Diego de Vargas and the Reconquest of New Mexico*. Albuquerque: University of New Mexico Press.

Bancroft, G., and R. B. Nye

1966 *The History of the United States of America from the Discovery of the Continent*. Chicago: University of Chicago Press.

Bandelier, A. F. A.

1885 The Romantic School in American Archaeology. Paper read before the New York Historical Society, February 3, New York.

1881 *Papers of the Archaeological Institute of America*. American Series I–IV. Boston: Williams and Trübner.

1887–1888 *Histoire de la Colonisation et des Missions de Sonora, Chihuahua, Nouveau Mexique, et Arizona jusqu'a l'Année 1700*. Santa Fe, NM: N.p.

1890 *Final Report of Investigations among the Indians of the Southwestern United States, Carried on Mainly in the Years from 1880 to 1885: Volume I*. Cambridge, MA: Wilson.

1892 *Final Report of Investigations among the Indians of the Southwestern United States, Carried on Mainly in the Years from 1880 to 1885: Volume II*. Cambridge, MA: Wilson.

1893 *The Gilded Man (El Dorado) and Other Pictures of the Spanish Occupancy of America*. New York: Appleton.

1910 *Documentary History of the Rio Grande Pueblos of New Mexico*. Santa Fe, NM: n.p.

1971 *The Delight Makers*. New York: Harcourt Brace Jovanovich.

Bandelier, A. F. A., and Archaeological Institute of America

1881 *Historical Introduction to Studies among the Sedentary Indians of New Mexico: Report on the Ruins of the Pueblo of Pecos*. Papers of the

Archaeological Institute of America. American Series vol. 1. Boston: Williams and Trübner.

Bandelier, A. F. A., and G. P. Hammond

1949 *A Scientist on the Trail: Travel Letters of A. F. Bandelier, 1880–1881.* Berkeley, CA: Quivira Society.

Bandelier, A. F. A., C. H. Lange, C. L. Riley, and E. M. Lange

1966 *The Southwestern Journals of Adolph F. Bandelier: 1880–1882.* Albuquerque: University of New Mexico Press.

1970 *The Southwestern Journals of Adolph F. Bandelier: 1883–1884.* Albuquerque: University of New Mexico Press.

1975 *The Southwestern Journals of Adolph F. Bandelier: 1885–1888.* Albuquerque: University of New Mexico Press.

1984 *The Southwestern Journals of Adolph F. Bandelier: 1889–1893.* Albuquerque: University of New Mexico Press.

Bandelier, A. F. A., L. H. Morgan, and L. A. White

1978 *Pioneers in American Anthropology: The Bandelier-Morgan Letters, 1873–1883.* New York: AMS Press.

Banks, Marcus

1996 *Ethnicity: Anthropological Constructions.* New York: Routledge Press.

Bannon, J. F.

1966 *Indian Labor in the Spanish Indies: Was There Another Solution?* Problems in Latin American Civilization. Boston: Heath.

1970 *The Spanish Borderlands Frontier, 1513–1821.* Histories of the American Frontier. New York: Holt, Rinehart, and Winston.

1978 *Herbert Eugene Bolton: The Historian and the Man, 1870–1953.* Tucson: University of Arizona Press.

Barrett, Elenor M.

1997 *Conquest and Catastrophe: Changing Rio Grande Pueblo Settlement Pattern in the Sixteenth and Seventeenth Centuries.* Albuquerque: University of New Mexico Press.

2002 The Geography of the Rio Grande Pueblos in the Seventeenth Century. *Ethnohistory* 49(1):123–169.

Barth, Fredrick

1969 Introduction. In *Ethnic Groups and Boundaries.* Oslo: Norwegian University Press.

1994 Enduring and Emerging Issues in the Analysis of Ethnicity. In *The Anthropology of Ethnicity*, edited by Hans Vermeulen and Cora Govers. Amsterdam: Het Spinhuis.

Barth, Fredrick, and Universitetet i Bergen

1969 *Ethnic Groups and Boundaries: The Social Organization of Culture Difference.* Results of a symposium held at the University of Bergen, 23 to 26 February 1967. Oslo: Scandinavian University Books; Bergen: Universitetsforlaget; London: Allen and Unwin.

Bayer, L., and F. Montoya

1994 *Santa Ana: The People, the Pueblo, and the History of Tamaya.* Albuquerque: University of New Mexico Press.

Benavides, A. de, F. W. Hodge, A. Rey, and G. P. Hammond

1945 *Fray Alonso de Benavides' Revised Memorial of 1634: With Numerous Supplementary Documents Elaborately Annotated.* Coronado Cuarto Centennial Publications, 1540–1940, Vol. IV. Albuquerque: University of New Mexico Press.

Benedict, R.

1931 *Tales of the Cochiti Indians.* Bulletin 98. Washington, DC: Smithsonian Institution Press.

Bentley, C. G.

1987 Ethnicity and Practice. *Comparative Studies in Society and History* 29:24–55.

Berry, M. S.

1985 The Age of Maize in the Greater Southwest: A Critical Review. In *Prehistoric Food Production in North America*, edited by R. Ford. Anthropological Papers No. 75, Museum of Anthropology, University of Michigan. Ann Arbor: University of Michigan Press.

Bhabha, H. K.

1990 *Nation and Narration.* London: Routledge Press.

Binford, Lewis A.

1962 Archeology as Anthropology. *American Antiquity* 28(2):217–225.

1964 A Consideration of Archaeological Research Design. *American Antiquity* 29(4):425–441.

1967 Reply to K. C. Chang's "Major Aspects of the Interrelationship of Archaeology and Ethnology." *Current Anthropology* 8 (3):234–235.

1968a Methodological Considerations of the Archeological Use of Ethnographic Data. In *Man the Hunter*, edited by R. B. L. and I. Devore. Chicago: University of Chicago Press.

1968b Some Comments on Historical versus Processual Archeology. *Southwestern Journal of Anthropology* 24(3):267–275.

Blakey, M. L.

1990 American Nationality and Ethnicity in the Depicted Past. In *The Politics of the Past*, edited by Peter Gathercole and David Lowenthal. London: Unwin Hyman.

Bloom, L. B.

1922 The West Jemez Culture Area. *El Palacio* 12:18–25.

1931 A Campaign against the Moqui Pueblos under Gobernador Phelix Martinez, 1716. *New Mexico Historical Review* 5:158–226.

1938 The Chapter Elections in 1672. *New Mexico Historical Review* 13(85–119).

Bolton, H. E.

1930 *Spanish Exploration in the Southwest, 1542–1706*. New York: Scribner.

1936 *Rim of Christendom: A Biography of Eusebio Francisco Kino, Pacific Coast Pioneer*. New York: Russell and Russell.

1949 *Coronado, Knight of Pueblos and Plains*. New York: Whittlesey House.

1990 *Coronado, Knight of Pueblos and Plains*. 4th ed. Albuquerque: University of New Mexico Press.

Bolton, H. E., and J. F. Bannon

1964 *Bolton and the Spanish Borderlands*. Norman: University of Oklahoma Press.

Bolton, H. E., and H. B. Foster

1915 *Explorations on the Northern Frontier of New Spain 1535–1706*. Berkeley: University of California Press.

Bolton, H. E., and T. M. Marshall

1919 *The Colonization of North America, 1492–1783*. (Authors' preliminary copy, with corrections in ms.) New York: Macmillan.

Bolton, H. E., and University of New Mexico Center for the American West.

1996 *The Spanish Borderlands: A Chronicle of Old Florida and the Southwest*. Historians of the Frontier and American West. Albuquerque: University of New Mexico Press.

Borah, W. W.

1951 *New Spain's "Century of Depression."* Ibero-Americana 35. Berkeley: University of California Press.

Borah, W. W., and S. F. Cook

1963 The Aboriginal Population of Central Mexico on the Eve of the Spanish Conquest. *Ibero-Americana* 45. Berkeley: University of California Press.

Bourdieu, P.

1977 *Outline of a Theory of Practice*. Cambridge: Cambridge University Press.

Brass, P. R.

1985 *Ethnic Groups and the State*. London: Croom Helm.

Brew, J. O.

1937 *The First Two Seasons at Awatovi*. Menasha, WI: N.p.

Bromley, Y. V.

1981 *Sovremennye Problemy Ethnografii*. Moscow: N.p.

Bunzell, Ruth

1938 The Economic Organization of Primitive Peoples. Pp. 327–408 in *General Anthropology*, edited by Franz Boas. Boston: Heath.

Burrus, E. J.

1969 *A History of the Southwest: A Study of the Civilization and Conversion of the Indians in Southwestern United States and Northwestern Mexico from the Earliest Times to 1700. Supplement to Volume I. Reproduction in Color of Thirty Sketches and of Ten Maps*. Vatican City: Biblioteca Apostolica Vaticana.

Carbia, R.

1943 *Historia de la Leyenda Negra Hispano-Americana.* Buenos Aires: N.p.

Carlson, I. K., and T. A. Kohler

1990 *Prolegomenon to the Study of Habitation Site Architecture during the Coalition Period on the Pjarito Plateau.* Reports of Investigations 62. Pullman: Department of Anthropology, Washington State University.

Carlson, R. L.

1965 *Eighteenth Century Navajo Fortresses in the Gobernador District.* University of Colorado Studies Series in Anthropology 10. Boulder: University of Colorado Press.

Carr, E. H.

1961 *What Is History?* New York: Vintage Books.

Casas, B. de las, and M. J. Zúñiga y Ontiveros

1552 *Breve relación de la destrucción de las Indias occidentales: presentada a Felipe II siendo príncipe de Asturias por don Fray Bartolomé de las Casas.* Mexico City: Mariano Ontiveros.

Castañeda de Nájera, P. de, F. Vázquez de Coronado, A. de Mendoza, J. C. Jaramillo, A. Parker, and G. P. Winship

1933 *The Journey of Francisco Vazquez de Coronado, 1540–1542.* San Francisco: Grabhorn Press.

Chapman, K. M.

1911 *Notes on the Burial Mounds and Rooms; Amoxiumqua, 1911. Notes on file at USFS.* Santa Fe, NM: U.S. Forest Service, Santa Fe National Forest.

Chilton, E. S.

1996 *Embodiments of Choice: Native American Ceramic Diversity in the New England Interior.* Amherst: University of Massachusetts Press.

1998 The Cultural Origins of Technological Choice: Unraveling Algonquin and Iriquoian Ceramic Traditions in the Northeast. In *The Archaeology of Social Boundaries*, edited by M. Stark. Washington, DC: Smithsonian Institution Press.

Chipman, D. E.

1987 In Search of Cabeza de Vaca's Route across Texas: An Historiographical Survey. *Southwestern Historical Quarterly* 91 (October):127–148.

Cohen, Abner

1969 *Custom and Politics in Urban Africa: A Study of Hausa Migrants in Yoruba Towns.* London: Routledge Press.

Cohen, Anthony P.

1985 *The Symbolic Construction of Community.* London: Routledge Press.

Cohen, Ronald

1978 Ethnicity: Problem and Focus in Anthropology. In *Annual Review of Anthropology.* Palo Alto, CA: Annual Reviews.

Cohn, B. S.

1987 *An Anthropologist among the Historians and Other Essays*. New York: Oxford University Press.

Collingwood, R. G.

1946 *The Idea of History*. Oxford: Clarendon Press.

Colwell-Chanthaphonh, C., and T. J. Ferguson

2008 *Collaboration in Archaeological Practice: Engaging Descendant Communities*. Lanham, MD: AltaMira Press.

Comaroff, Jean

1985 *Body of Power, Spirit of Resistance: The Culture and History of a South African People*. Chicago: University of Chicago Press.

Comaroff, Jean, and John L. Comaroff

1991 *Of Revelation and Revolution*. Chicago: University of Chicago Press.

Comaroff, John L.

1987 Of Totemism and Ethnicity: Consciousness, Practice and Signs of Inequality. *Ethnos* 52(3–4):301–323.

Comaroff, John L., and American Bar Foundation

1994 *Ethnicity, Nationalism, and the Politics of Identity in an Age of Revolution*. ABF Working Paper no. 9402. Chicago: American Bar Foundation.

Conkey, M. W., and C. A. Hastorf (editors)

1990 *The Uses of Style in Archaeology*. Cambridge: University of Cambridge Press.

Conway, G. R. G.

1931 Antonio de Espejo as a Familiar of the Mexican Inquisition, 1572–1578. *New Mexico Historical Review* 6:1–20.

Cook, David Noble, and W. George Lovell (editors)

1981 *"Secret Judgments of God": Old World Disease in Colonial Spanish America*. Norman: University of Oklahoma Press.

Cook, S. F.

1976 *The Conflict between the California Indian and White Civilization*. Berkeley: University of California Press.

Cook, S. F., and L. B. Simpson

1948 The Population of Central Mexico in the Sixteenth Century. *Ibero-Americana* 31. Berkeley: University of California Press.

Cordell, Linda S.

1977 Late Anasazi Farming and Hunting Strategies. *American Antiquity* 42:449–461.

1996 Big Sites, Big Questions: Pueblos in Transition. In *The Prehistoric Pueblo World: A.D. 1150–1350*, edited by M. A. Adler. University of Arizona Press, Tucson.

1997 *Archaeology of the Southwest.* 2nd ed. San Diego, CA: Academic Press.

1984 *Prehistory of the Southwest.* New York: Academic Press.

Cordell, Linda S., David E. Doyel, and Keith W. Kintigh

1994 Processes of Aggregation in the Prehistoric Southwest. In *Themes in Southwest Prehistory*, edited by G. S. Gummerman. Santa Fe, NM: School of American Research Press.

Cordell, Linda S., and Amy C. Earls

1984 The Rio Grande Glaze "Sequence" and the Mogollon. In *Recent Research in Mogollon Archaeology*, edited by S. Upham, F. Plog, D. Batcho, and B. E. Kauffman. New Mexico State University Museum Occasional Papers No. 10. Las Cruces: New Mexico State University.

Cordell, L. S., G. J. Gumerman, and School of American Research (Santa Fe, NM)

1989 *Dynamics of Southwest Prehistory.* Smithsonian Series in Archaeological Inquiry. Washington, DC: Smithsonian Institution Press.

Cordell, Linda S., and School of American Research (Santa Fe, NM)

1984 *Prehistory of the Southwest.* New World Archaeological Record. Orlando, FL: Academic Press.

Cordell, Linda S.; United States Bureau of Land Management, Albuquerque District; United States Forest Service, Southwestern Region; and United States Bureau of Land Management, New Mexico State Office

1979 *A Cultural Resources Overview of the Middle Rio Grande Valley, New Mexico.* USDA Forest Service Southwestern Region, Bureau of Land Management New Mexico State Office. Washington, DC: Government Printing Office.

Cordell, Linda S., and Vincent J. Yannie

1991 Ethnicity, Ethnogenesis and the Individual: A Processual Approach toward Dialogue. In *Processual and Postprocessual Archaeologies: Multiple Ways of Knowing the Past*, edited by R. W. Preucel, vol. 10, pp. 96–107. Carbondale: Southern Illinois University Press.

Creamer, W.

1993 *The Architecture of Arroyo Hondo Pueblo, New Mexico.* Santa Fe, NM: School of American Research Press

1994 Re-examining the Black Legend: Contact Period Demography in the Rio Grande Valley of New Mexico. *New Mexico Historical Review* 69(3):263–278.

Creamer, W., C. M. Cameron, and J. D. Beal

1993 *The Architecture of Arroyo Hondo Pueblo, New Mexico.* Arroyo Hondo Archaeological Series, vol. 7. Santa Fe, NM: School of American Research Press

Creamer, W., and L. Renken

1994 Testing Conventional Wisdom: Protohistoric Ceramics and Chronology in the Northern Rio Grande. Paper Presented at the 59th

Annual Meeting of the Society of American Archaeology, April 22, Anaheim, CA.

Crosby, A. W.

1987 *The Columbian Voyages, the Columbian Exchange, and Their Historians: Essays on Global and Comparative History.* Washington, DC: American Historical Association.

Crown, P. L., W. J. Judge, and School of American Research (Santa Fe, NM)

1991 *Chaco and Hohokam: Prehistoric Regional Systems in the American Southwest.* School of American Research Advanced Seminar series. Santa Fe, NM: School of American Research Press.

Crown, P. L., and T. A. Kohler

1994 Community Dynamics, Site Structure, and Aggregation in the Northern Rio Grande. In *The Ancient Southwestern Community: Models and Methods for the Study of Prehistoric Social Organization*, edited by W. H. and R. D. L. Wills. Albuquerque: University of New Mexico Press.

Crown, P. L., J. D. Orcutt, and T. A. Kohler

1996 Pueblo Cultures in Transition: The Northern Rio Grande. In *The Pueblo World, 1150–1300*, edited by M. A. Adler, pp. 188–204. Tucson: University of Arizona Press.

Cushing, Frank H.

1920 *Zuni Breadstuff.* Indian Notes and Monographs 8. New York: Museum of the American Indian, Heye Foundation.

Darwin, C.

1871 *The Descent of Man, and Selection in Relation to Sex.* London: Murray.

Davis, D. B.

1966 *The Problem of Slavery in Western Culture.* Ithaca, NY: Cornell University Press.

Dean, J. S.

1988 A Model of Anasazi Behavioral Adaptation. In *The Anasazi in a Changing Environment*, edited by G. J. Gummerman, pp. 25–44. Cambridge: Cambridge University Press.

Dean, J. S., Robert C. Euler, George J. Gummerman, Fred Plog, Richard Hevly, and Thor Karlstrom

1985 Human Behavior, Demography and Paleoenvironment on the Colorado Plateaus. *American Antiquity* 50:537–554.

Deeds, S. M.

1998 First Generation Rebellions in Seventeenth Century Nueva Vizcaya. In *Native Resistance and the Pax Colonial in New Spain*, edited by S. Schroeder, pp. 1–29. Lincoln: University of Nebraska Press.

Diamond, J. M.

1997 *Guns, Germs, and Steel: The Fates of Human Societies.* New York: Norton.

Dirks, N. B.

1987 *The Hollow Crown: Ethnohistory of an Indian Kingdom*. Cambridge South Asian Studies. Cambridge: Cambridge University Press.

1992 *Colonialism and Culture*. Comparative Studies in Society and History. Ann Arbor: University of Michigan Press.

Dobyns, H. F.

1963 *An Outline of Andean Epidemic History to 1720*. Baltimore, MD: Johns Hopkins University Institute of the History of Medicine.

1976 *Native American Historical Demography: A Critical Bibliography*. Bloomington: Indiana University Press.

Dobyns, H. F.; W. R. Swagerty; and Newberry Library, Center for the History of the American Indian

1983 *Their Number Become Thinned: Native American Population Dynamics in Eastern North America*. Native American Historic Demography series. Knoxville: University of Tennessee Press in cooperation with the Newberry Library Center for the History of the American Indian.

Dougherty, Julia D.

1980 *Refugee Pueblos on the Santa Fe National Forest*. Cultural Resources Report 2. Santa Fe, NM: Santa Fe National Forest.

Dougherty, Julia D., and David P. Staley

1979 *An Archaeological Investigation of Horn Mesa (Southwestern End) and Kotyiti Pueblo*. USDA Forest Service Report, Santa Fe Forest, Tesuque Ranger District, 1979-10-035.

Doyel, D.

1976 Classic Period Hohokam in the Gila Basin. *The Kiva* 42:27–38.

Dozier, E. P.

1952 *The Changing Social Organization of the Hopi-Tewa*. PhD diss., University of California, Los Angeles.

1954 *The Hopi-Tewa of Arizona*. University of California Publications in American Archaeology and Ethnology, vol. 44, no.3. Berkeley: University of California Press.

1955 Forced and Permissive Acculturation. *American Indian* 7I(2):38–44.

1962 Differing Reactions to Religious Contacts among North American Indian Societies. In *Acts of the 34th International American Congresses*, pp. 161–171. Vienna: N.p.

1966 *Hano, a Tewa Indian Community in Arizona*. New York: Holt, Rinehart, and Winston.

1970a Making Inferences from the Present to the Past. In *Reconstructing Prehistoric Pueblo Societies* pp. 202–213. Albuquerque: University of New Mexico Press.

1970b *The Pueblo Indians of North America*. Case Studies in Cultural Anthropology. New York: Holt, Rinehart, and Winston.

Dumarest, N.

1919 Notes on Cochiti New Mexico. *Memoirs of the American Anthropological Association* 6(3):137–236.

Eggan, F.

1950 *Social Organization of the Western Pueblos*. Chicago: University of Chicago Press.

Elliott, J. H.

1969 *Imperial Spain 1469–1716*. London: Penguin Books.

1989 *Spain and Its World*. New Haven, CT: Yale University Press.

1998 *The Old World and the New: 1492–1650*. Cambridge: Cambridge University Press.

Elliott, M. L.

1982 *Large Pueblo Sites Near Jemez Springs, New Mexico*. Cultural Resources Report Number 3. Washington, DC: U.S. Forest Service.

1993 *Jemez*. Santa Fe: Museum of New Mexico Press.

Ellis, F. H.

1964 *A Reconstruction of the Basic Jemez Pattern of Social Organization, with Comparisons to Other Tanoan Social Structures*. Albuquerque: University of New Mexico Press.

Ellis, F. H., and Florence Hawley Ellis Museum of Anthropology

1989 *San Gabriel del Yungue as Seen by an Archaeologist*. Santa Fe, NM: Sunstone Press in conjunction with the Florence Hawley Ellis Museum of Anthropology at Ghost Ranch, Abiquiu, New Mexico.

Ellis, F. H., and United States Indian Claims Commission

1974 *Archaeologic and Ethnologic Data: Acoma-Laguna Land Claims*. Pueblo Indians 2. New York: Garland.

Elmore, F. H.

1976 *Shrubs and Trees of the Southwest Uplands*. Globe, AZ: Southwest Parks and Monuments Association.

Eriksen, T. H.

1993 *Ethnicity and Nationalism: Anthropological Perspectives*. Anthropology, Culture, and Society. London: Pluto Press.

Espinosa, J. Manuel

1942 *Crusaders of the Rio Grande*. Chicago: Institute of Jesuit History.

1988 *The Pueblo Revolt of 1696 and the Franciscan Missions in New Mexico: Letters of the Missionaries and Related Documents*. Norman: University of Oklahoma Press.

Euler, R. C., George J. Gummerman, Thor N. V. Karlstrom, Jeffrey S. Dean, and Richard H. Hevly

1979 The Colorado Plateaus: Cultural Dynamics and Paleoenvironment. *Science* 205:1089–1101.

Evans-Pritchard, E. E.
1940 The Nuer of the Southern Sudan. In *African Political Systems*, edited by Meyer Fortes and E. E. Evans-Pritchard. London: Oxford University Press.

Everson, Stephen (editor)
1996 *Aristotle: The Politics and the Constitution of Athens.* Cambridge: Cambridge University Press.

Fabian, J.
1983 *Time and the Other: How Anthropology Makes Its Object.* New York: Columbia University Press.

Farriss, N. M.
1978 Nucleation versus Dispersal: The Dynamics of Population Movement in Colonial Yucatan. *Hispanic American Historical Review* 58(2): 187–216.

1984 *Maya Society under Colonial Rule: The Collective Enterprise of Survival.* Princeton, NJ: Princeton University Press.

Ferguson, T. J.
1996 *Historic Zuni Architecture and Society: An Archaeological Application of Space Syntax.* Anthropological Papers of the University of Arizona, no. 60. Tucson: University of Arizona Press.

Ferguson, T. J., and C. Colwell-Chanthaphonh
2003 *History Is in the Land: Multivocal Tribal Traditions in Arizona's San Pedro Valley.* Tucson: University of Arizona Press.

Ferguson, T. J., E. R. Hart, T. Lucio, and R. Stauber
1985 *A Zuni Atlas.* Norman: University of Oklahoma Press.

Ferguson, T. J., and B. J. Mills
1982 *Archaeological Investigations at Zuni Pueblo: 1977–1980.* Zuni Archaeology Program Report 183. Pueblo of Zuni, New Mexico.

Ferguson, T. J., and R. W. Preucel
2005 Signs of the Ancestors: An Archaeology of the Mesa Villages of the Pueblo Revolt. In *Structure and Meaning in Human Settlement*, edited by T. Atkin and J. Rykwert, pp. 185–207. Philadelphia: University Museum Press.

Fewkes, J. W.
1898 Archaeological Expedition into Arizona in 1895. *Seventeenth Annual Report of the Bureau of American Ethnology for the Years 1895–1896* 2:519–742.

Firth, R. W., and British Academy
1972 *The Sceptical Anthropologist, Social Anthropology and Marxist Views on Society.* London: Oxford University Press for the British Academy.

Fish, P. R. S. K. F., George J. Gummerman, and J. Jefferson Reid
1994 Toward an Explanation of Abandonments. In *Themes in Southwest Prehistory*, edited by G. J. Gummerman. Santa Fe, NM: School of American Research Press.

Folsom, F.
1973 *Indian Uprising on the Rio Grande: The Pueblo Revolt of 1680*. Albuquerque: University of New Mexico Press.

Forbes, Jack D.
1950 Unknown Athapaskans: The Identification of the Jano, Jacome, Jumano, Manso, Suma, and Other Indian Tribes of the Southwest. *Ethnology* 6(2):97–159.
1960 *Apache, Navaho, and Spaniard*. Norman: University of Oklahoma Press.
1963 *The Historian and the Indian: Racial Bias in American History*. Washington, DC: Academy of American Franciscan History.

Ford, James A.
1952 Measurements of Some Prehistoric Design Elements in the Southeastern States. *Anthropological Papers of the American Museum of Natural History* 44(3).

Ford, R.
1975 Re-excavation of Jemez Cave, New Mexico. *Awanyu* 3(3):13–27.

Foucault, M.
1970 *The Order of Things: An Archaeology of the Human Sciences*. London: Tavistock.

Foucault, M., and C. Gordon
1980 *Power Knowledge: Selected Interviews and Other Writings, 1972–1977*. Brighton, U.K.: Harvester Press.

Fuentes, C.
1992 *The Buried Mirror: Reflections on Spain and the New World*. Boston: Houghton Mifflin.

Furnivall, J. S.
1948 *Colonial Policy and Practice: A Comparative Study of Burma and Netherlands India*. Cambridge: Cambridge University Press.

Gachupin, J.
1993 *Site Inspection Report for Kotyiti (FS 1440)*. Santa Fe National Forest, Jemez Ranger District.

Gallegos Lamero, H., G. P. Hammond, and A. Rey
1927 *The Gallegos Relation of the Rodriguez Expedition to New Mexico*. Santa Fe, NM: El Palacio Press.

Gellner, Ernest
1983 *Nations and Nationalism*. New Perspectives on the Past. Ithaca, NY: Cornell University Press.
1987 *Culture, Identity and Politics*. Cambridge: Cambridge University Press.

Gerhard, Peter
1981 *The North Frontier of New Spain*. Princeton, NJ: Princeton University Press.

Gero, J. M., D. M. Lacy, and M. L. Blakey (editors)

1983 *The Socio-politics of Archaeology*. Research Report 23, Department of Anthropology. Amherst: University of Massachusetts.

Gibson, C.

1958 *Spain in America*. New York: Harper and Row.

1964 *The Aztecs under Spanish Rule: A History of the Indians of the Valley of Mexico, 1519–1810*. Stanford, CA: Stanford University Press.

Giddens, A.

1984 *The Constitution of Society: Outline of the Theory of Structuration*. Cambridge: Polity Press.

1987 *A Contemporary Critique of Historical Materialsim: Vol. 2. The Nation State and Violence*. Berkeley: University of California Press.

Giudicelli, Christophe, and Pierre Ragon

2000 Les martyrs ou la Vierge? Frères martyrs et images outragées dans le Mexique du Nord (XVIème–XVIIème siècles). *Cahiers des Amériques Latines* (2nd series), 33(1):32–55. Paris: Institut des Hautes Études de l'Amérique latine (IHEAL), Université Paris III—Sorbonne Nouvelle and Centre de Recherche et de Documentation sur l'Amérique Latine.

Glassow, M. A.

1980 *Prehistoric Agricultural Development in the Northern Southwest: A Study in Changing Patterns of Land Use*. Anthropological Papers No. 16. Socorro, NM: Ballena Press.

Glazer, Nathan, and Daniel P. Moynihan

1975 *Ethnicity: Theory and Experience*. Cambridge, MA: Harvard University Press.

Gluckman, Max

1940 The Kingdom of the Zulu in South Africa. In *African Political Systems*, edited by M. Fortes and E. E. Evans-Pritchard. London: Oxford University Press.

1958 *Analysis of a Social Situation in Modern Zululand*. Manchester: Manchester University Press for the Rhodes-Livingstone Institute.

Goff, F., J. N. Gardner, G. Valentine, New Mexico Bureau of Mines and Mineral Resources, and Los Alamos National Laboratory

1990 *Geology of St. Peter's Dome Area, Jemez Mountains, New Mexico*. Socorro: New Mexico Bureau of Mines and Mineral Resources.

Goldfrank, Esther Schiff

1927 The Social and Ceremonial Organization of Cochiti. *Memoirs of the American Anthropological Association* no. 33.

Goldstein, L.

1998 Editor's Corner. *American Antiquity* 63(3):373–374.

Gosden, Chris

2001 Postcolonial Archaeology: Issues of Culture, Identity and Knowledge. In *Archaeological Theory Today*, edited by I. Hodder. Cambridge: Polity Press.

Hammond, George P., and Agapito Rey (editors)

1928 *Obregon's History of Sixteenth Century Explorations in Western America.* Los Angeles: Wetzel.

Hanke, L.

1938 *The Requerimiento and Its Interpreters.* Mexico City: Imperial Mundial.

1941 *The Development of Regulations for Conquistadores.* Buenos Aires: Talleres and Casa J. Peuser.

1970 *Aristotle and the American Indians: A Study in Race Prejudice in the Modern World.* Bloomington: Indiana University Press.

1979 *Selected Writings of Lewis Hanke on the History of Latin America.* Tempe: Center for Latin American Studies Arizona State University.

Hanke, L., and B. de las Casas

1974 *All Mankind Is One: A Study of the Disputation between Bartolomé de Las Casas and Juan Ginés de Sepúlveda in 1550 on the Intellectual and Religious Capacity of the American Indians.* DeKalb: Northern Illinois University Press.

Hannaford, Ivan

1996 *Race: The History of an Idea in the West.* Baltimore, MD: Johns Hopkins University Press.

Harlow, F. H.

1973 *Matte Paint Pottery of the Tewa, Keres and Zuni Pueblos.* Albuquerque: Museum of New Mexico.

Harrington, John P.

1916 The Ethnogeography of the Tewa Indians. In *Twenty-ninth Annual Report of the Bureau of American Ethnology.* Washington, DC: Government Printing Office.

Harris, M.

2000 *Aztecs, Moors, and Christians: Festivals of Reconquest in Mexico and Spain.* Austin: University of Texas Press.

Hayes, A. C., and T. J. Caperton

1981 *Contributions to Gran Quivira Archeology, Gran Quivira National Monument, New Mexico.* Washington, DC: National Park Service, U.S. Department of the Interior.

Hayes, A. C., J. N. Young, and A. H. Warren

1981 *Excavation of Mound 7, Gran Quivira National Monument, New Mexico.* National Park Service, U.S. Department of Interior. Washington, DC: Government Printing Office.

Henige, D.

1978 On the Contact Population of Hispaniola: History as Higher Mathematics. *Hispanic American Historical Review* 58(2):217–237.

1986 If Pigs Could Fly: Timucan Population and Native American Historical Demography. *Journal of Interdisciplinary History* 16(4):700–721.

Hester, J. J.

1962 *Early Navajo Migrations and Acculturation in the Southwest.* Museum of New Mexico Papers in Anthropology 6. Albuquerque: Museum of New Mexico

Hewett, E. L.

1906 . . . *Antiquities of Jemez Plateau, New Mexico; by Edgar L. Hewett.* Washington, DC: Government Printing Office.

Hewett, E. L., and R. G. Fisher

1943 *Mission Monuments of New Mexico.* Albuquerque: University of New Mexico Press.

Hewett, E. L., S. G. Morley, and K. M. Chapman

1938 *Pajarito Plateau and Its Ancient People.* Albuquerque: University of New Mexico Press.

Hill, James N.

1970 *Broken K Pueblo: Prehistoric Social Organization in the American Southwest.* Anthropological Papers of the University of Arizona, No. 18. Tucson: University of Arizona Press.

Hill, James N., and W. Nicholas Trierweiler

1986 *Prehistoric Responses to Food Stress on the Pajarito Plateau, New Mexico: Technical Report and Results of the Pajarito Archaeology Research Project, 1977–1985*, University of California, Los Angeles.

Hill, W. W., and C. H. Lange

1982 *An Ethnography of Santa Clara Pueblo, New Mexico.* Albuquerque:

Hobsbawn, Eric J.

1983 Introduction: Inventing Traditions. In *The Invention of Tradition*, edited by E. Hobsbawm and T. Ranger, pp. 1–14. Cambridge: Cambridge University of New Mexico Press.

Hobsbawm, Eric J., and Terence O. Ranger

1983 *The Invention of Tradition.* Cambridge: Cambridge University Press.

Hodder, Ian

1983 Boundaries as Strategies. In *The Archaeology of Frontiers and Boundaries*, edited by S. W. Green and S. M. Perlman, pp. 141–159. Orlando, FL: Academic Press.

1985 Boundaries as Strategies: An Ethnoarchaeological Study. In *The Archaeology of Frontiers and Boundaries*, edited by S. W. Green and S. M. Perlman. New York: Academic Press.

1986 *Reading the Past: Current Approaches to Interpretation in Archaeology.* Cambridge: Cambridge University Press.

1992 *Theory and Practice in Archaeology: Material Cultures.* London: Routledge Press.

1995 *Interpreting Archaeology: Finding Meaning in the Past.* London: Routledge Press.

Hodge, Frederick Webb

1896 *Pueblo Indian Clans.* Washington, DC: Judd and Detweiler.

1910 Handbook of American Indians: Part II. *Bureau of American Ethnology Bulletin* 30:210.

1916 *Memorial of Fray Alonzo de Benavides, 1630.* Edited by E. E. Ayer, pp. 242–243, note 32. Chicago: Private printing.

1937 *History of Hawikuh, New Mexico: One of the So-called Cities of Cíbola.* Publications, vol. 1. Los Angeles: Southwest Museum.

1954 *Tówayálane: A Mesa with a History.* N.p.

Hodge, Frederick Webb, George P. Hammond, and Agapito Rey

1945 *Fray Alonso de Benavides' Revised Memorial of 1634.* Albuquerque: University of New Mexico Press.

Hodge, Frederick Webb, W. Smith, R. B. Woodbury, N. F. S. Woodbury, and Hendricks-Hodge Expedition

1918 *The Excavation of Hawikuh by Frederick Webb Hodge: Report of the Hendricks-Hodge Expedition, 1917–1923.* New York: Museum of the American Indian, Heye Foundation.

Holmes, W. H.

1905 Notes on the Antiquities of the Jemez Valley, New Mexico. *American Anthropologist* 7:198–212.

Horowitz, Donald L.

1975 Ethnic Identity. In *Ethnicity: Theory and Experience,* edited by Nathan Glazer and Daniel P. Moynihan. Cambridge, MA: Harvard University Press.

1977 Cultural Movements and Ethnic Change. In *Ethnic Conflict in the World Today: The Annals of the American Academy of Political and Social Science,* edited by Richard D. Lambert and Alan W. Heston, 433:1–19.

1983 *Ethnic Groups in Conflict.* Los Angeles: University of California Press.

1985 *Ethnic Groups in Conflict.* 2nd ed. Berkeley: University of California Press.

Housley, L.

1974 *Opuntia Imbricata Distribution on Old Jemez Indian Habitation Sites.* Master's thesis, Claremont College.

Hurt, W. R., and U.S. National Park Service Division of History and Southwest Cultural Resources Center, Division of Anthropology

1990 *The 1939–1940 Excavation Project at Quarai Pueblo and Mission Buildings: Salinas Pueblo Missions National Monument, New Mexico.* Southwest Cultural Resources Center Professional Papers no. 29. Santa

Fe, NM: Division of History, Division of Anthropology National Park Service.

Hurtado, A.

1993 Herbert E. Bolton, Racism and American History. *Pacific Historical Review* 62 (May):127–142.

1995 Parkmanizing the Spanish Borderlands: Bolton, Turner and the Historian's World. *Western Historical Quarterly* 26 (Summer):149–167.

Instituto Nacional para el Federalismo y el Desarrollo Municipal (INAFED)

2005 Nochistlán de Mejía, Zazatecas. In *Enciclopedia de los Municipios de México* (online edition at E-Local). Mexico City: INAFED, Secretaría de Gobernación, Gobierno del Estado de Zacatecas.

Isaac, Barry C.

1993 Retrospective on the Formalist-Substantivist Debate. *Research in Economic Anthropology* 13:213–233.

Isaacs, Harold R.

1975 Basic Group Identity: The Idols of the Tribe. In *Ethnicity: Theory and Experience*, edited by Nathan Glazer and Daniel P. Moynihan, pp. 29–53. Cambridge, MA: Harvard University Press.

Ivey, J. E.

1998 Convento Kivas in the Missions of New Mexico. *New Mexico Historical Review* (April):121–152.

Jackson, R. H.

1982 *Demographic and Social Change in Northwestern New Spain: A Comparative Analysis of the Pimería Alta and Baja California Missions.*

1994 *Indian Population Decline: The Missions of Northwestern New Spain, 1687–1840*. Albuquerque: University of New Mexico Press.

1998 *New Views of Borderlands History*. Albuquerque: University of New Mexico Press.

1999 *Race, Caste, and Status: Indians in Colonial Spanish America*. Albuquerque: University of New Mexico Press.

Jackson, R. H., and E. D. Castillo

1995 *Indians, Franciscans, and Spanish Colonization: The Impact of the Mission System on California Indians*. Albuquerque: University of New Mexico Press.

Jenkins, R.

1996 *Social Identity: Key Ideas*. London: Routledge Press.

Jones, O.

1966 *Pueblo Warriors and Spanish Conquest*. Norman: University of Oklahoma Press.

Jones, S.

1997 *The Archaeology of Ethnicity: Constructing Identities in the Past and Present*. London: Routledge Press.

Juderias, J.
1917 *La Leyenda Negra*. 4th ed. Barcelona: N.p.

Kamen, H. A. F.
1984 *European Society 1500–1700*. London: Hutchison.
1991 *Spain, 1469–1714: A Society of Conflict*. 2nd ed. New York: Longman.
1998 *The Spanish Inquisition: A Historical Revision*. New Haven, CT: Yale University Press.

Keen, B.
1985 Main Currents in United States Writings on Colonial Spanish America, 1884–1984. *Hispanic American Historical Review* 65(4):657–682.

Kelley, J. Charles
1950 The Historic Indian Pueblos of La Junta de los Ríos. *New Mexico Historical Review* 27:257–295, 28:21–51.

Kessell, John L.
1979 *Kiva, Cross, and Crown: The Pecos Indians and New Mexico, 1540–1840*. Washington, DC: National Park Service, U.S. Department of the Interior.
1989 Spaniards and Pueblos: From Crusading Intolerance to Pragmatic Accommodation. In *Columbian Consequences*, vol. 1, edited by D. H. Thomas. Washington, DC: Smithsonian Institution Press.

Kessell, John L., and R. Hendricks
1991 *The Spanish Missions of New Mexico*. Spanish Borderlands Sourcebooks 17–18. New York: Garland.

Kessell, John L., Rick Hendricks, and Meredith Dodge (editors)
1995 *To the Royal Crown Restored: The Journals of Don Diego de Vargas, 1692–1694*. Albuquerque: University of New Mexico Press.

Kidder, A. V.
1916 *The Pueblo of Pecos*. Papers of the School of American Archaeology, no. 33. Washington, DC: School of American Archaeology.
1924 *An Introduction to the Study of Southwestern Archaeology with a Preliminary Account of the Excavations at Pecos*. Papers of the Philips Academy Southwestern Expedition, no. 1. New Haven, CT: Yale University Press.
1927 Southwestern Archaeological Conference. *Science* 64:489–491.
1958 *Pecos, New Mexico: Archaeological Notes*. Papers of the Robert S. Peabody Foundation for Archaeology, vol. 5. Andover, MA: Phillips Academy.
1964 *Pottery of the Pajarito Plateau and of Some Adjacent Regions in New Mexico*. New York: Kraus Reprint.

Kidder, A. V., and Robert S. Peabody Foundation for Archaeology
1924 *An Introduction to the Study of Southwestern Archaeology with a Preliminary Account of the Excavations at Pecos*. Department of

1993b *Technical Choices: Transformation in Material Cultures since the Neolithic*. London: Routledge Press.

Leroi-Gourhan, André, and Anna Bostocl Berger (translator)
1993 *Le geste et la parole* (Gesture and Speech). Cambridge, MA: MIT Press.

Lewis, B.
1995 *Cultures in Conflict: Christians, Muslims, and Jews in the Age of Discovery*. New York: Oxford University Press.

Liebman, Matthew
2005 *Burn the Churches, Break Up the Bells: The Archaeology of the Pueblo Revolt in New Mexico, A.D. 1680–1696*. PhD diss., University of Pennsylvania.

Liebman, Matthew, T. J. Ferguson, and R. W. Preucel
2005 Pueblo Settlement, Architecture, and Social Change in the Pueblo Revolt Era, A.D. 1680–1696. *Journal of Field Archaeology* 30:1–16.

Lightfoot, D. R.
1993 The Landscape Context of Anasazi Pebble-Mulched Fields in the Galisteo Basin, Northern New Mexico. *Geoarchaeology* 8:349–370.

1994 The Agricultural Utility of Lithic Mulch Gardens: Past and Present. *GeoJournal* 34(4):425–437.

1995 The Construction and Configuration of Anasazi Period Pebble-Mulch Gardens in the Northern Rio Grande. *American Antiquity* 60(3): 459–470.

Lightfoot, Kent G.
2005 Indians, Missionaries, and Merchants: the Legacy of Colonial Encounters on the California Frontiers. Berkeley, CA: University of California Press.

Lightfoot, Kent, and Antoinette Martinez
1993 Frontiers and Boundaries in Archaeological Perspective. *Annual Review of Anthropology* 24:471–492.

1993 Culture Contact Studies: The Relationship between Prehistoric and Historical Archaeology. *American Antiquity* 60:199–218.

Lintz, C.
1991 Texas Panhandle-Pueblo Interactions from the Thirteenth through the Sixteenth Century. In *Farmers, Hunters and Colonists: Interaction between the Southwest and Southeastern Plains*, edited by K. A. Spielmann. Tucson: University of Arizona Press.

Longacre, William A.
1968 *Archaeology as Anthropology: A Case Study*. Anthropological Papers of the University of Arizona 17. Tucson: University of Arizona Press.

López-Portillo y Weber, José
1939 *La rebelión de Nueva Galicia, Pan American Institute of Geography and History, Publication 37*. Tacabuya, Mexico: Murguia.

Lowenthal, D.
1985 *The Past Is a Foreign Country*. Cambridge: Cambridge University Press.

Lummis, C. F.

1909 *The Spanish Pioneers.* Chicago: Rio Grande Press.

1956 *Los conquistadores españoles del siglo XVI; vindicación de la acción colonizadora española en América.* Guatemala City: Editora Latino Americana.

Lycett, M. T.

1995 *Archaeological Implications of European Contact: Demography, Settlement, and Land Use in the Middle Rio Grande Valley.* Albuquerque: University of New Mexico.

Mackenney, R.

1993 *Sixteenth Century Europe: Expansion and Conflict.* New York: St. Martin's Press.

MacLachlan, Colin M.

1988 *Spain's Empire in the New World: The Role of Ideas in Institutional and Social Change.* Berkeley: University of California Press.

Magnaghi, Russell M.

1998 *Herbert E. Bolton and the Historiography of the Americas.* Westport, CT: Greenwood Press.

Maltby, W. S.

1971 *The Black Legend in England: The Development of Anti-Spanish Sentiment, 1558–1660.* Duke Historical Publications. Durham, NC: Duke University Press.

Marcus, G. E., and M. M. J. Fischer

1986 *Anthropology as Cultural Critique: An Experimental Moment in the Human Sciences.* Chicago: University of Chicago Press.

Martin, C.

1987 *The American Indian and the Problem of History.* New York: Oxford University Press.

Masse, W. Bruce

1991 The Quest for Subsistence Sufficiency and Civilization in the Sonoran Desert in Chaco and Hohokam. In *Prehistoric Regional Systems in the American Southwest,* edited by P. L. Crown and W. J. Judge, pp. 195–224. Santa Fe, NM: School of American Research Press.

McGhee, Robert

2008 Aboriginalism and the Problems of Indigenous Archaeology. *American Antiquity* 73(4):579–597.

McNiven, Ian J., and Lynette Russell

2005 *Appropriated Pasts: Indigenous Peoples and the Colonial Culture of Archaeology.* Walnut Creek, CA: AltaMira Press.

McWilliams, C.

1946 *Southern California Country: An Island on the Land.* New York: Books for Libraries Press.

Mead, G. H., and C. W. Morris

1934 *Mind, Self and Society from the Standpoint of a Social Behaviorist.* Chicago: University of Chicago Press.

Mecham, J. L.

1924a The Second Spanish Expedition to New Mexico. *New Mexico Historical Review* 1:266–267.

1924b Supplementary Documents Relating to the Chamuscado-Rodríguez Expedition. *Southwestern Historical Quarterly* 29(1):224–231.

Meltzer, D. J.

1979 Paradigms and the Nature of Change in American Archaeology. *American Antiquity* 44:644–657.

1983 The Antiquity of Man and the Development of American Archaeology. *Advances in Archaeological Method and Theory* 6:1–51.

Mera, H. P.

1933 *A Proposed Revision of the Rio Grande Glaze Paint Sequence.* Santa Fe, NM: Archaeological Survey Laboratory of Anthropology.

1934 *A Survey of the Biscuit Ware Area in Northern New Mexico.* Santa Fe, NM: Laboratory of Anthropology.

1939a *Style Trends of Pueblo Potters in the Rio Grande and Little Colorado Cultural Areas, from the Sixteenth to the Nineteenth Century.* Memoirs of the Laboratory of Anthropology, Santa Fe III. Santa Fe, NM: Laboratory of Anthropology.

1939b *Style Trends of Pueblo Pottery in the Rio Grande and Little Colorado Cultural Areas from the Sixteenth to the Nineteenth Century.* Santa Fe, NM: Waverly Press.

1940a *Ceramic Clues to the Prehistory of North Central New Mexico.* Technical Series Bulletin no. 8. Ann Arbor, MI: Edwards.

1940b *Population Changes in the Rio Grande Glaze-Paint Area.* Laboratory of Anthropology Technical Series, Bulletin No. 9. Santa Fe, NM: Archaeological Survey Laboratory of Anthropology.

1991 *Style Trends of Pueblo Pottery, 1500–1840.* Albuquerque: Avanyu.

Meskell, Lynn

2005 Archaeological Ethnography: Conversations around Kruger National Park. *Archaeologies.* Volume 1, number 1. Pp 81–100. New York: Springer.

Meskell, L., and R. W. Preucel

2004 *A Companion to Social Archaeology.* Malden, MA: Blackwell.

Meyer, M. C., and W. L. Sherman

1991 *The Course of Mexican History.* 4th ed. New York: Oxford University Press.

Meyer, M. C., W. L. Sherman, and S. M. Deeds

1999 *The Course of Mexican History.* 6th ed. New York: Oxford University Press.

Miller, R. R.

1985 *Mexico: A History.* Norman: University of Oklahoma Press.

Mills, Barbara J. (editor)
2004 *Identity, Feasting, and the Archaeology of the Greater Southwest.* Boulder, CO: University Press of Colorado.

Mills, B. J., and P. L. Crown
1995 *Ceramic Production in the American Southwest.* Tucson: University of Arizona Press.

Mindeleff, Cosmos
1900 Localization of the Tusayan Clans. In *Eighth Annual Report of the Bureau of American Ethnology for the Years 1886–1887*, pp. 635–653. Washington, DC: Bureau of American Ethnology.

Mindeleff, Victor
1891 A Study of Pueblo Architecture at Tusayan and Cibola. In *Eighth Annual Report of the Bureau of American Ethnology for the Years 1886–1887*, pp. 3–228. Washington, DC: Bureau of American Ethnology.

Minnis, P. E.
1985 Domesticating People and Plants in the Greater Southwest. In *Prehistoric Food Production in North America*, edited by R. Ford, pp. 309–339. Museum of Anthropology Papers No. 75. Ann Arbor: University of Michigan Press.

Minnis, P. E., and C. L. Redman
1990 *Perspectives on Southwestern Prehistory.* Investigations in American Archaeology. Boulder, CO: Westview Press.

Mitchell, J. Clyde
1959 *The Kalela Dance: Aspects of Social Relationships among Urban Africans in Northern Rhodesia.* Published on behalf of the Rhodes-Livingstone Institute. Manchester: Manchester University Press.

Monroy Castillo, María Isabel, and Tomás Calvillo Unna
1997 *Breve historia de San Luis Potosí.* Reproduced online at the Biblioteca Digital, Instituto Latinoamericano de la Comunicación Educativa (ILCE), Serie breves historias de los estados de la República Mexicana. Mexico City: Colegio de México, ideicomiso Historia de las Américas, and Fondo de Cultura Económica.

Montgomery, R. G.
1949 *Franciscan Awatovi: The Excavation and Conjectural Reconstruction of a Seventeenth-Century Spanish Mission Establishment at a Hopi Indian Town in Northeastern Arizona.* Cambridge, MA: Peabody Museum of Archaeology and Ethnology, Harvard University.

Montoya, J. de, G. P. Hammond, and A. Rey
1938 *New Mexico in 1602: Juan de Montoya's Relation of the Discovery of New Mexico.* Albuquerque: Quivira Society.

Moore, J. L.
1994 Prehistoric Pueblo Farming in the Taos District. In *Artifacts, Shrines and Pueblos: Papers in Honor of Gordon Paige*, edited by Melina S. Duran and D. T. Kirkpatrick. Albuquerque: Archaeological Society of New Mexico.

Morgan, L. H.

1877 *Ancient Society: Researches in the Lines of Human Progress from Savagery through Barbarism to Civilization.* Tucson: University of Arizona Press.

Narroll, R.

1963 On Ethnic Unit Classification, *Current Anthropology* 5:283–291.

Naylor, T. H., and C. W. Polzer

1986 *The Presidio and Militia on the Northern Frontier of New Spain: A Documentary History.* Tucson: University of Arizona Press.

Nelson, N. C.

1912 *Correspondence to Charles Wissler.* New York: Archives of the American Museum of Natural History.

1914 *Pueblo Ruins of the Galisteo Basin, New Mexico.* New York: Archives of the American Museum of Natural History.

1916 Chronology of the Tano Ruins, New Mexico. *American Anthropologist* 18:159–180.

n.d. *Pueblo Kotyiti: Field Notes.* New York: Archives of the American Museum of Natural History.

n.d. Excavations of Pueblo Kotyiti, New Mexico. New York: Archives of the American Museum of Natural History.

Nicholas, George P., and T. D. Andrews

1997 *At a Crossroads: Archaeology and First Peoples in Canada.* Burnaby, BC: Archaeology Press.

Obregón, B. de, G. P. Hammond, and A. Rey

1928 *Obregón's History of 16th Century Explorations in Western America: Entitled Chronicle, Commentary, or Relation of the Ancient and Modern Discoveries in New Spain and New Mexico, Mexico, 1584.* Los Angeles: Wetzel Press.

Olsen, S. J., and R. P. Wheeler

1978 *Bones from Awatovi, Northeastern Arizona: The Faunal Analysis.* Report no. 11. Cambridge, MA: Peabody Museum of Archaeology and Ethnology, Harvard University.

Ong, A.

1987 *Spirits of Resistance and Capitalist Discipline: Factory Women in Malaysia.* SUNY Series in the Anthropology of Work. Albany: State University of New York Press.

Ortiz, Alfonso

1969 *The Tewa World: Space Time and Becoming in a Pueblo Society.* Chicago: University of Chicago Press.

1979a *Handbook of North American Indians*: Vol. 9. Washington, DC: Smithsonian Institution Press.

1979b Po'pay's Leadership: A Pueblo Perspective. *El Palacio* 86(4):11–33 (Museum of New Mexico).

Oviedo y Valdes, G. F. de, B. C. Hedrick, and C. L. Riley
1974 *The Journey of the Vaca Party: The Account of the Narvaez Expedition, 1528–1536, as Related by Gonzalo Fernandez de Oviedo y Valdes.* Cabondale: Southern Illinois University Press.

Pagden, Anthony
1993 *European Encounters with the New World: From Renaissance to Romanticism.* New Haven, CT: Yale University Press.

Palkovich, A. M.
1985 Historic Population of the Eastern Pueblos: 1540–1910. *Journal of Anthropological Research* 41(4):401–426.

Park, Robert E.
1921 Sociology and the Social Sciences: The Group Concept and Social Research. *American Journal of Sociology* 27(1):169–183.

1950 *Race and Culture.* Glencoe, IL: Free Press.

1955 *Society: Collective Behavior, News and Opinion, Sociology and Modern Society.* Glencoe, IL: Free Press.

Park, Robert E., and Ernest W. Burgess
1924 *Introduction to the Science of Sociology.* Chicago: University of Chicago Press.

Parkman, F.
1865 *France and England in North America: A Series of Historical Narratives.* Boston: Little, Brown.

1879 *La Salle and the Discovery of the Great West.* 11th rev. ed. Boston: Little, Brown.

1904 *The Oregon Trail.* Boston: Little, Brown.

Parsons, E. C.
1939 *Pueblo Indian Religion.* Vols. 1–2. Chicago: University of Chicago Press.

Patterson, O.
1982 *Slavery and Social Death: A Comparative Study.* Cambridge, MA: Harvard University Press.

Pérez de Luxán, D., G. P. Hammond, and A. Rey
1929 *Expedition into New Mexico Made by Antonio de Espejo, 1582–1583, as Revealed in the Journal of Diego Pérez de Luxán, a Member of the Party.* Los Angeles: Quivira Society.

1967 *Expedition into New Mexico Made by Antonio de Espejo, 1582–1583, as Revealed in the Journal of Diego Pérez de Luxán, a Member of the Party.* New York: Arno Press.

Phelan, J. L.
1980 *The Millennial Kingdom of the Franciscans in the New World: A Study of the Writings of Gerónimo de Mendieta (1525–1604).* Millwood, NY: Kraus Reprint.

Plog, F.

1978 Social Interaction and Stylistic Similarity: A Reanalysis. In *Advances in Archaeological Method and Theory*, edited by M. B. Schiffer, vol. 1, pp. 143–182. New York: Academic Press.

1980 *Stylistic Variation in Prehistoric Ceramics: Design Analysis in the Prehistoric Southwest.* Cambridge: Cambridge University Press.

1983 Analysis of Style in Artifacts. *Annual Review of Anthropology* 12:125–142.

1990 Sociopolitical Implications of Stylistic Variation in the American Southwest. In *The Uses of Style in Archaeology*, edited by M. Conkey and C. Hastorf, pp. 61–72. Cambridge: Cambridge University Press.

1995 Approaches to Style: Compliments and Contrasts. In *Style, Society and Person: Archaeological and Ethnological Perspectives*, edited by C. Carr and J. E. Neitzel. New York: Plenum Press.

Plog, F., George J. Gummerman, Robert C. Euler, Jeffrey, S. Dean, Richard H. Hevly, and Thor N. V. Karlstrom

1988 Anasazi Adaptive Strategies: The Mode Prediction and Results. In *The Anasazi in a Changing Environment*, edited by G. J. Gummerman. Cambridge: Cambridge University Press.

Polzer, C. W.

1998 *Kino: His Missions, His Monuments.* Tucson: Jesuit Fathers of Southern Arizona.

Powell, P. W.

1953 *Soldiers, Indians and Silver: The Northward Advance of Spain, 1550–1600.* Berkeley: University of California Press.

Preucel, R. W.

1987 Settlement Succession on the Pajarito Plateau, New Mexico. *The Kiva* 53(1):3–34.

1990 *Seasonal Circulation and Dual Residence in the Pueblo Southwest: A Prehistoric Example from the Pajarito Plateau, New Mexico.* New York: Garland.

1991 Introduction. In *Processual and Postprocessual Archaeologies*, edited by R. W. Preucel, pp. 1–29. Carbondale: Southern Illinois University at Carbondale, Center for Anthropological Investigations.

1998 *The Kotyiti Research Project: Report of the 1996 Fieldseason.* Santa Fe, NM: U.S. Forest Service, Southwestern District.

2000a Living on the Mesa: Hanat Kotyiti's Post-Revolt Cochiti Community in the Northern Rio Grande. *Expedition* 42:8–17.

2000b Making Pueblo Identities: Architectural Discourse at Kotyiti, New Mexico. In *An Archaeology of Communities in the Americas*, edited by J. Yaeger and M. Canuto. London: Routledge Press.

2002 *Archaeologies of the Pueblo Revolt: Identity, Meaning, and Renewal in the Pueblo World.* Albuquerque: University of New Mexico Press.

2005a The Journey from Shipap. In *The Peopling of Bandelier: New Insights from the Archeology of the Pajarito Plateau*, edited by Robert P. Powers, pp. 95–101. Santa Fe, NM: School of American Research Press.

2005b Ethnicity and Southwestern archaeology. In *Southwestern Archaeology in the Twentieth Century*, edited by Linda Cordell and Don Fowler, pp. 174–193. Salt Lake City: University of Utah Press.

2006 *Archaeological Semiotics.* Oxford: Blackwell.

Preucel, R. W., and I. Hodder

1996 *Contemporary Archaeology in Theory.* Oxford: Blackwell.

Preucel, R. W., L. P. Traxler, and M. V. Wilcox

2002 "Now the God of the Spaniards Is Dead": Ethnogenesis and Community Formation in the Aftermath of the Pueblo Revolt of 1680. In *Traditions, Transitions and Technologies: Themes in Southwestern Archaeology*, edited by S. H. Schlanger. Boulder: University Press of Colorado.

Rabasa, José

2000 *Writing Violence on the Northern Frontier: The Historiography of Sixteenth-Century New Mexico and Florida and the Legacy of Conquest.* Durham: Duke University Press.

Raish, C.

1990 *Cultural Resources Survey for Fencing Seshukwa.* U.S. Forest Service, Jemez Ranger District.

1991 *Site Inspection Report for Kotyiti (FS 1440).* Santa Fe National Forest, Jemez Ranger District.

1993 *Site Inspection Report for Kotyiti (FS 1440).* Santa Fe National Forest, Jemez Ranger District.

Ramenofsky, Ann F.

1987 *Vectors of Death.* Albuquerque: University of New Mexico Press.

2005 *Agricultural Intensification on the Jemez Plateau: From Coalition to Early Historic.* Paper presented at the Seventy-first Annual Meeting of the Society for American Archaeology. Symposium on "Exploring the Archaeology of the Jemez Province."

Ramenofsky, Ann F., Shawn L. Penman, and Anastasia Steffen

1996 *Stable Mesa Ruin Mapping Project, University of New Mexico.* Santa Fe National Forest Report 1996-10-077.

Ramenofsky, Ann F, with C. D. Vaughan

1996 On the Question of Infectious Disease in New Mexico, 1540–1680. *Journal of Anthropological Research* 52:161–184.

Rautman, A. E.

1993 Resource Variability, Risk and the Structure of Social Networks: An Example from the Prehistoric Southwest. *American Antiquity* 58:403–424.

Reagan, A. B.

1922 Additional Notes on the Jemez-Zia Region. *El Palacio* 12:120–121.

Redman, C. L.

1991 In Defense of the Seventies: The Adolescence of the New Archaeology. *American Anthropologist* 93:295–307.

Reed, Erik K.

1943 The Southern Tewa Pueblos in the Historic Period. *El Palacio* 50:250–288.

Reff, D. T.

1991 *Disease, Depopulation, and Culture Change in Northwestern New Spain, 1518–1764*. Salt Lake City: University of Utah Press.

1998 The Jesuit Mission Frontier in Comparative Perspective: The Reductions of the Rio de la Plata and the Missions of northwestern Mexico, 1588–1700. In *Contested Ground: Comparative Frontiers on the Northern and Southern Edges of the Spanish Empire*, edited by D. Reff. Tucson: University of Arizona Press.

Reichard, Gladys A.

1938 Social Life. In *General Anthropology*, edited by Franz Boas. Boston: Heath.

Reiter, P.

1938 *The Jemez Pueblo of Unshagi*. Vols. 1 and 2. Albuquerque: University of New Mexico Press.

Reiter, P., William T. Mulloy, and E. H. Blumenthal

1940 Preliminary Report on the Jemez Excavation at Nanishagi, New Mexico. *University of New Mexico Bulletin, Anthropological Series* 3(3).

Rex, J., and D. Mason

1986 *Theories of Race and Ethnic Relations: Comparative Ethnic and Race Relations.* Cambridge: Cambridge University Press.

Ricard, R.

1974 *The Spiritual Conquest of Mexico: An Essay on the Apostolate and the Evangelizing Methods of the Mendicant Orders in New Spain, 1523–1572.* California Library Reprint series. Berkeley: University of California Press.

Riley, C. L.

1971 Early Spanish-Indian Interaction Communication in the Greater Southwest. *New Mexico Historical Review* 46(4):285–314.

1976 *Sixteenth Century Trade in the Greater Southwest.* Mesoamerican Studies no. 10. Carbondale: University Museum and Art Galleries, Southern Illinois University.

1987 *The Frontier People: The Greater Southwest in the Protohistoric Period.* Albuquerque: University of New Mexico Press.

1996 *Rio del Norte: People of the Upper Rio Grande from Earliest Times to the Pueblo Revolt.* Salt Lake City: University of Utah Press.

1997 The Teya Indians of the Southwestern Plains. In *The Coronado Expedition to Tierra Nueva: The 1540–1542 Route across the Southwest*, edited by R. and S. C. F. Flint, pp. 320–344. Boulder: University of Colorado Press.

Robinson, W. J., J. W. Hannah, and B. G. Harrill

1972 *Tree-Ring Dates from New Mexico: I, O, U, Central Rio Grande Area.* Tucson: Laboratory of Tree Ring Research, University of Arizona.

Rogers, J. D.

1990 *Objects of Change: The Archaeology and History of Arikara Contact with Europeans.* Washington, DC: Smithsonian Institution Press.

Rogers, J. D., and S. M. Wilson (editors)

1993 *Ethnohistory and Archaeology: Approaches to Postcontact Change in the Americas.* New York: Plenum Press.

Rojas, Beatriz; Jesús Gómez Serrano, Andrés Reyes Rodríguez, Salvador Camacho, and Carlos Reyes Sahagún

1994 *Breve historia de Aguascalientes.* Reproduced online at the Biblioteca Digital, Instituto Latinoamericano de la Comunicación Educativa (ILCE), Serie breves historias de los estados de la República Mexicana. Mexico City: Colegio de México, Fideicomiso Historia de las Américas, and Fondo de Cultura Económica.

Rubiano, P. Alvarez

1941 Importancia político-social de las mercedes de 1519 concedidas a los labradores de Tierra Firme. *Revista de Indias.*

Sachet, J. R.

1977 The Meaning of Style in Archaeology: A General Model. *American Antiquity* 43(3):369–380.

Sackett, J. R.

1982 Approaches to Style in Lithic Archaeology. *Journal of Anthropological Archaeology* 1(1):59–112.

1985a Style and Ethnicity in Archaeology: A Reply to Wiessner. *American Antiquity* 50(1):154–159.

1985b Style, Ethnicity and Stone Tools. In *Status Structure and Stratification: Current Archaeological Reconstructions*, edited by M. Thompson, M. T. Garcia, and F. Kearse. Calgary, Alberta: University of Calgary, Archaeological Association.

1986 Isochrestism and Style: A Clarification. *Journal of Anthropological Archaeology* 5:266–277.

1990 *Style and Ethnicity in Archaeology.* Cambridge: Cambridge University Press.

Said, E. W.

1978 *Orientalism.* New York: Pantheon Books.

1993 *Culture and Imperialism.* New York: Knopf.

Salmón, R. M.

1977 Tarahumara Resistance to Mission Congregation in Northern New Spain, 1580–1710. *Ethnohistory* 24(4):379–393.

Sando, J. S.

1992 *Pueblo Nations: Eight Centuries of Pueblo Indian History*. Santa Fe, NM: Clear Light Publishers.

1998 *Pueblo Profiles: Cultural Identity through Centuries of Change*. Santa Fe, NM: Clear Light Publishers.

Sando, Joe, and Herman Agoyo

2005 *Po'pay: Leader of the First American Revolution*. Santa Fe, NM: Clear Light Publishers.

Sauer, C. O.

1934 The Distribution of Aboriginal Tribes and Languages in Northwestern Mexico. *Ibero-Americana* 5. Berkeley: University of California Press.

1935 Aboriginal Population of Northwestern Mexico. *Ibero-Americana* 10. Berkeley: University of California Press.

1948 Colima of New Spain in the Sixteenth Century. *Ibero-Americana* 29. Berkeley: University of California Press.

Schiffer, Michael B.

1976 Introduction and a Synthetic Model of Inference. In *Behavioral Archaeology*, pp. 2–25. New York: Academic Press.

Schmal, John P.

2004 *The History of Zazatecas*. History of Mexico. Houston, TX: Houston Institute for Culture.

Schneider, J., and R. Rapp

1995 *Articulating Hidden Histories: Exploring the Influence of Eric R. Wolf*. Berkeley: University of California Press.

Scholes, Frances V.

1930 The Supply Service of the New Mexican Missions in the Seventeenth Century. *New Mexico Historical Review* 5:93–404.

1932 *Problems in the Early Ecclesiastical History of New Mexico*. Santa Fe: Historical Society of New Mexico.

1935a *Civil Government and Society in New Mexico in the Seventeenth Century*. N.p.

1935b *The First Decade of the Inquisition in New Mexico*. Santa Fe, NM: N.p.

1936 Church and State in New Mexico, 1610–1650. *New Mexico Historical Review* 11: 9–76ff.

1940 Church and State in New Mexico. *New Mexico Historical Review* 15: 78–106.

1942a *Church and State in New Mexico in the Seventeenth Century*. Albuquerque: University of New Mexico Press.

1942b *Troublous Times in New Mexico, 1659–1670*. Albuquerque: University of New Mexico Press.

Scholes, F. V., and H. P. Mera

1940 Some Aspects of the Jumano Problem. *Contributions to American Anthropology and History*, vol. 6, no. 34, pub. 523. Washington, DC: Carnegie Institute of Washington Publications.

Schroeder, Albert H.

1968 Shifting for Survival in the Spanish Southwest. *New Mexico Historical Review* 43:291–310.

1975 Rio Grande Ethnohistory. In *New Perspectives on the Pueblos*, edited by A. Ortiz, pp. 41–70. Albuquerque: University of New Mexico Press.

1978 Pueblos Abandoned in Historic Times. In *Handbook of North American Indians*, edited by Alfonso Ortiz, pp. 236–254. Washington, DC: Smithsonian Institution Press.

1979 Pueblos Abandoned in Historic Times. In *Handbook of North American Indians, Vol. 9: Southwest*, edited by A. Ortiz, pp. 236–254. Washington, DC: Smithsonian Institution Press.

1979 Shifting for Survival in the Spanish Southwest. In *New Spain's Far Northern Frontier: Essays on Spain in the American West, 1540–1821*, edited by D. J. Weber, pp. 239–255. Albuquerque: University of New Mexico Press.

Schroeder, Susan (editor)

1996 *Native Resistance and the Pax Colonial of New Spain*. Lincoln: University of Nebraska Press.

Scott, J. C.

1985 *Weapons of the Weak: Everyday Forms of Peasant Resistance*. New Haven, CT: Yale University Press.

Seed, P.

1995 *Ceremonies of Possession in Europe's Conquest of the New World, 1492–1640*. New York: Cambridge University Press.

Shennan, Stephen

1989a *Archaeological Approaches to Cultural Identity*. One World Archaeology 10. London: Unwin Hyman.

1989b Introduction. In *Archaeological Approaches to Cultural Identity*, pp. 1–32. One World Archaeology. London: Unwin Hyman.

Shepherd, Nick

2002 Heading South, Looking North. Why We Need a Postcolonial Archaeology. *Archaeological Dialogues* 9(2).

Shirokokogoroff, S.

1923 *Etnos, Issledovanie osnovnykh printsipov etnicheskikh I etnograficheskikh yavlenii*, Shanghai: N.p.

Silliman, Stephen (editor)

2008 *Collaborating at the Trowel's Edge: Teaching and Learning in Indigenous Archaeology*. Amerind Studies in Archaeology no. 2. Tucson: University of Arizona Press.

Silverberg, R.

1970 *The Pueblo Revolt*. New York: Weybright and Talley.

Simmons, A. H.

1986 New Evidence for the Early Use of Cultigens in the American Southwest. *American Antiquity* 51:73–89.

Simpson, L. B.

1929 *The* Encomienda *in New Spain: Forced Native Labor in the Spanish Colonies, 1492–1550*. Berkeley: University of California Press.

1934 *Studies in the Administration of the Indians in New Spain*. Berkeley: University of California Press.

1952 Exploitation of Land in Central Mexico in the Sixteenth Century. *Ibero-Americana* 36. Berkeley: University of California Press.

Smiley, T. H., Stanley A. Stubbs, and Bryant Bannister

1953 A Foundation for the Dating of Some Later Archaeological Sites in the Rio Grande Area, New Mexico. *Laboratory of Tree Ring Research, Bulletin 6*. Tucson: Laboratory of Tree Ring Research, University of Arizona.

Smith, Anthony D.

1986 *The Ethnic Origins of Nations*. Oxford: Blackwell.

Smith, B., G. P. Hammond, M. Dean, and Book Club of California

1946 *The Discovery of Florida, Being a True Relation of the Vicissitudes That Attended the Governor Don Hernando de Soto and Some Nobles of Portugal in the Discovery of Florida Now Just Given by a Didalgo of Elvas*. San Francisco: Grabhorn Press for the Book Club of California.

Smith, Claire, and H. Martin Wobst (editors)

2005 *Indigenous Archaeologies: Decolonizing Theory and Practice (One World Archaeology)*. New York: Routledge Press.

Smith, W.

1952 *Kiva Mural Decorations at Awatovi and Kawaika-a, with a Survey of Other Wall Paintings in the Pueblo Southwest*. Reports of the Awatovi Expedition, Report no. 5. Cambridge, MA: Peabody Museum of Archaeology and Ethnology, Harvard University.

1971 *Painted Ceramics of the Western Mound at Awatovi*. Report no. 8. Cambridge, MA: Peabody Museum of Archaeology and Ethnology, Harvard University.

Smith, W., R. B. Woodbury, and N. S. F. Woodbury

1966 *The Excavation of Hawikuk, by Frederick Webb Hodge*. Contributions from the Museum of the American Indian Heye Foundation XX. New York: Heye Foundation.

Snead, James, and R. W. Preucel
1999 The Ideology of Settlement: Ancestral Keres Landscapes in the Northern Rio Grande. In *Archaeologies of Landscape: Contemporary Perspectives*, edited by W. Ashmore and A. B. Knapp. Oxford: Blackwell.

Sokolovskii, Sergey, and Valery Tishkov
1996 Ethnicity. In *Encyclopedia of Social and Cultural Anthropology*, edited by Alan Barnard and Jonathan Spencer. London: Routledge Press.

Sollors, Werner
1992 *Theories of Ethnicity: A Classical Reader.* New York: New York University Press.

Spicer, Edward H.
1962 *Cycles of Conquest: The Impact of Spain, Mexico, and the United States on the Indians of the Southwest, 1533–1795*. Tucson: University of Arizona Press.

Spielman, K. A.
1983 Late Prehistoric Exchange between the Southwest and Southern Plains. *Plains Anthropologist* 28:257–272.

1991a Interaction among Nonhierarchical Societies. In *Farmers, Hunters and Colonists: Interaction between the Southwest and Southern Plains*, edited by K. A. Spielman. Tucson: University of Arizona Press.

1991b *Interdependence in the Prehistoric Southwest: An ecological Analysis of Plains-Pueblo Interaction*. New York: Garland.

1996 Impressions of Pueblo II Settlement Trends Among the Rio Abajo and Eastern Border Pueblos. In *The Prehistoric Pueblo World: A.D. 1150–1350*, edited by M. A. Adler. Tucson: University of Arizona Press.

Spielmann, K. A., Margaret Schoeninger, and Katherine Moore
1990 Plains-Pueblo Interdependence and Human Diet at Pecos Pueblo, New Mexico. *American Antiquity* 55(4):745–765.

Spivak, G. C., D. Landry, and G. M. MacLean
1996 *The Spivak Reader: Selected Works of Gayatri Chakravorty Spivak*. New York: Routledge Press.

Stallings, W. S. J.
1933 A Tree Ring Chronology for the Rio Gande Drainage in Northern New Mexico. *Proceedings: National Academy of Sciences* 19:803–806.

1937 Southwest Dated Ruins. *Tree Ring Bulletin* 4(2). Laboratory of Tree Ring Research, University of Arizona, Tucson, Arizona.

Stark, M. T.
1998 *The Archaeology of Social Boundaries.* Washington, DC: Smithsonian Institution Press.

Steflen, A.
1995 *Site Inspection Report for Kotyiti (FS 1440)*. Santa Fe National Forest, Jemez Ranger District.

Steward, Julian
1955 *Theory of Culture Change*. Champaign: University of Illinois Press.

Sullivan, F.
1995 *Indian Freedom: The Cause of Bartolomé de las Casas, 1484–1566: A Reader*. Kansas City, MO: Sheed and Ward.

Taylor, W. B.
1979 *Drinking, Homicide and Rebellion in Colonial Mexican Villages*. Stanford, CA: Stanford University Press.

Taylor, W. W.
1948 A Study of Archaeology. *American Anthropological Association Memoir* 69.

Thomas, David Hurst (editor)
1989 *Columbian Consequences*. Vols. 1–3. Washington, DC: Smithsonian Institution Press.

Thomas, David Hurst
2007 *American Archaeology in the Twenty-First Century: Back to the Future?* In T.W. Killion (editor), Opening Archaeology: Repatriation's Impact on Contemporary Research and Practice, 57–78. Santa Fe, NM: School of American Research Press.

Thomas, David Hurst, and Society for American Archaeology
1989 *Columbian Consequences: Volume 1. Archaeological and Historical Perspectives on the Spanish Borderlands West*. Washington, DC: Smithsonian Institution Press.

Toulouse, J. H.
1949 *The Mission of San Gregorio de Abó: A Report on the Excavation and Repair of a Seventeenth-Century New Mexico Mission*. Albuquerque: University of New Mexico Press.

Toulouse, J. H., and R. L. Stephanson
1960 *Excavations at Pueblo Pardo, Central New Mexico*. Museum of New Mexico Papers in Anthropology No. 2. Santa Fe: Museum of New Mexico.

Trierweiler, W. N.
1990 *Prehistoric Tewa Economy: Modeling Subsistence and Production on the Pajarito Plateau*. New York: Garland.

Turner, F. J.
1996 *The Frontier in American History*. New York: Dover.

Twitchell, R. E.
1914 *The Spanish Archives of New Mexico; Comp. and Chronologically Arranged with Historical, Genealogical, Geographical, and Other Annotations, by Authority of the State of New Mexico*. Cedar Rapids, IA: Torch Press.

Udall, S. L., and J. D. Jacka
1987 *To the Inland Empire: Coronado and Our Spanish legacy*. New York: Doubleday Press.

Upham, S. (editor)
1987 *The Tyranny of Ethnographic Analogy in Southwestern Archaeology.* Anthropological Research Papers, No. 38, Tempe, Arizona.

Urban, Greg
1996 *Metaphysical Community: The Interplay of the Senses and the Intellect.* Austin: University of Texas Press.

van den Berghe, P. L.
1976 Race and Ethnicity: A Sociological Perspective. In *Ethnic and Racial Studies* 1(4):401–411.

1981 *The Ethnic Phenomenon.* London: Elsevier Press.

Vargas, D. de, R. Hendricks, M. D. Dodge, and J. L. Kessell
1992 *By Force of Arms: The Journals of Don Diego de Vargas, New Mexico, 1691–93.* Albuquerque: University of New Mexico Press.

Vargas, D. de, and J. L. Kessell
1989 *Remote beyond Compare: Letters of Don Diego de Vargas to His Family from New Spain and New Mexico, 1675–1706.* Albuquerque: University of New Mexico Press.

2000 *That Disturbances Cease: The Journals of Don Diego de Vargas, New Mexico, 1697–1700.* Albuquerque: University of New Mexico Press.

Vargas, D. de, J. L. Kessell, R. Hendricks, and M. D. Dodge
1992a *By Force of Arms: The Journals of Don Diego de Vargas, New Mexico, 1691–93.* Albuquerque: University of New Mexico Press.

1992b *Letters from the New World: Selected Correspondence of Don Diego de Vargas to His Family, 1675–1706.* Albuquerque: University of New Mexico Press.

1995 *To the Royal Crown Restored: The Journals of Don Diego de Vargas, New Mexico, 1692–94.* Works. Albuquerque: University of New Mexico Press.

1998 *Blood on the Boulders: The Journals of Don Diego de Vargas, New Mexico, 1694–97.* Albuquerque: University of New Mexico Press.

Vargas, D. de, and R. E. Twitchell
1916 *Pueblo Revolt of June 4, 1696.* Washington, DC: Archaeological Institute of America.

Verano, J. W., D. H. Ubelaker, and National Museum of Natural History (U.S.)
1992 *Disease and Demography in the Americas.* Washington, DC: Smithsonian Institution Press.

Vermeulen, H., and C. Govers
1994 *The Anthropology of Ethnicity: Beyond "Ethnic Groups and Boundaries."* Amsterdam: Spinhuis.

Villagrá, G. P. de, and G. Espinosa
1933 *History of New Mexico, Alcalá, 1610.* Los Angeles: Quivira Society.

Vivian, R. G.
1964 *Gran Quivira: Excavations in a Seventeenth-Century Jumano Pueblo.* Washington, DC: National Park Service, U.S. Department of the Interior.

1979 *Gran Quivira: Excavations in a Seventeenth-Century Jumano Pueblo.* Archeological Research Series, no. 8. Albuquerque, NM: National Park Service, U.S. Department of the Interior.

Warren, A. H.

1970 *Notes on File at the Museum of New Mexico, Laboratory of Anthropology.* Albuquerque: Museum of New Mexico.

1971 *Tamaya: A. R. M. File.* Albuquerque: Museum of New Mexico.

1979 The Glaze Paint Wares of the Upper Rio Grande Valley. In *Archaeological Investigations in Cochiti Reservoir, New Mexico: Volume 4: Adaptive Change in the Northern Rio Grande Valley*, edited by J. V. and R. C. C. Biella. Albuquerque: University of New Mexico Office of Contract Archaeology.

Warren, A. H., and D. H. Snow

1976 Formal Descriptions of Rio Grande Glazes from LA 70. *Laboratory of Anthropology Notes, Museum of New Mexico* 78:C1–C34.

Weber, David J.

1986 Turner, the Boltonians and the Borderlands. *American Historical Review* 91 (February):66–81.

1988 *Myth and the History of the Hispanic Southwest: Essays.* Calvin P. Horn Lectures in Western History and Culture. Albuquerque: University of New Mexico Press.

1992 *The Spanish Frontier in North America.* New Haven, CT: Yale University Press.

1999 *What Caused the Pueblo Revolt of 1680?* Boston: Bedford/St. Martin's.

Weber, David J., and Jane M. Rausch (editors)

1992 *Where Cultures Meet: Frontiers in Latin American History.* Wilmington, DE: Scholarly Resources.

Weber, Max

1922 Ethnic Groups. In *Theories of Ethnicity: A Classical Reader*, edited by Werner Sollors. Cambridge, MA: Harvard University Press.

Wendorf, F.

1953 *Excavations at Te'ewi.* Archaeological Institute of America, School of American Research Monographs, no. 17:34–93.

Wendorf, F., and E. K. Reed

1955 An Alternative Reconstruction of Northern Rio Grande Prehistory. *El Palacio* 62(5–6):131–173.

Werbner, R. P.

1984 The Manchester School in South-Central Africa. *Annual Review of Anthropology* 13:157–185.

White, L.

1940 *Pioneers in American Anthropology: The Bandelier-Morgan Letters, 1873–1883.* Vols. 1–2. Albuquerque: University of New Mexico Press.

1959 *The Evolution of Culture*. New York: McGraw-Hill.

1964 *The World of the Keresan Pueblo Indians in Primitive Views of the World*, edited by S. Diamond, pp. 83–94. New York: Columbia University Press.

Wiessner, P.

1983 Style and Social Information in Kalahari San Projectile Points. *American Antiquity* 48(2):253–276.

1984 Reconsidering the Behavioral Basis for Style: A Case Study among the Kalahari San. *Journal of Anthropological Archaeology* 3(3):190–234.

1985 Style or Isochrestic Variation? A Reply to Sackett. *American Antiquity* 50(1):160–166.

1990 Is There a Unity to Style? In *The Uses of Style in Archaeology*, edited by M. W. Conkey and C. A. Hastorf. Cambridge: Cambridge University Press.

Wilcox, D., and Jonathan Haas

1993 The Scream of the Butterfly: Competition and Conflict in the Prehistoric Southwest. In *Themes in Southwest Prehistory*. Santa Fe, NM: School of American Research Press.

Willey, G. R., and J. Sabloff

1991 *A History of American Archaeology*. New York: Freeman.

Williams, R. A.

1990 *The American Indian in Western Legal Thought*. Oxford: Oxford University Press.

Wills, W. H.

1990 Cultivating Ideas: The Changing Intellectual History of the Introduction of Agriculture in the American Southwest. In *Perspectives on Southwestern Prehistory*, edited by P. and C. R. Minnis. Boulder, CO: Westview Press.

Wills, W. H., and R. D. Leonard

1994 *The Ancient Southwestern Community: Models and Methods for the Study of Prehistoric Social Organization*. Albuquerque: University of New Mexico Press.

Wobst, H. M.

1977 Stylistic Behavior and Information Exchange. In *For the Director: Research Essays in Honor of James B. Griffin*, edited by C. E. Cleveland. Ann Arbor, Michigan: University of Michigan, Museum of Anthropology.

1978 The Archaeo-Ethnology of Hunter-Gatherers or The Tyranny of the Ethnographic Record in Archaeology. *American Antiquity* 43(2): 303–309.

Wolf, E. R.

1959 *Sons of the Shaking Earth*. Chicago: University of Chicago Press.

1966 *Sons of the Shaking Earth*. 2nd ed. Chicago: University of Chicago Press.

1982 *Europe and the People without History*. Berkeley: University of California Press.

Woosley, A. M.

1980 Agricultural Diversity in the Prehistoric Southwest. *Kiva* 43:317–335.

Worcester, D. E.

1991 Herbert Eugene Bolton: The Making of a Western Historian. In *Writing Western History: Essays on Major American Historians*, edited by R. W. Etulain. Albuquerque: University of New Mexico Press.

Wylie, Alison

1982 An Analogy by Any Other Name Is Just as Analogical. *Journal of Anthropological Archaeology* 1:382–401.

1987 "Simple" Analogy and the Role of Relevance Assumptions: Implications of Archaeological Practice. *International Studies in the Advancement of Science* 19:1–18.

1989 Archaeological Cables and Tacking: The Implications of Practice for Brenstein's "Options beyond Objectivism and Relativism." *Philosophy of Social Science* 19:1.

Zambardino, R. A.

1980 Mexico's Population in the Sixteenth Century: Demographic Anomaly or Mathematical Illusion? *Journal of Interdisciplinary History* 10(1) (Summer):1–27.

Zárate Salmerón, G. D., and J. F. Ramírez

1949 *Noticias sacadas de un mss. intitulado Relaciones de todas las cosas que en el Nuevo México se han visto y sabido, así por mar como por tierra desde el año 1538 hasta el de 1626*. Vargas Rea, Mexico: Biblioteca Aportación Histórica.

Zimmerman, J. F.

1940 The Coronado Cuarto Centennial. *Hispanic American Historical Review* 20: 158–162.

Zimmerman, Larry J.

1996 Epilogue: A New and Different Archaeology? *American Indian Quarterly: Special Issue: Repatriation: An Interdisciplinary Dialogue* 20(2): 297–307.

Index

Page numbers followed by *f* refer to figures, *n* refer to notes, and *t* refer to tables.